BUILDING AN AMERICAN CLOCK MOVEMENT

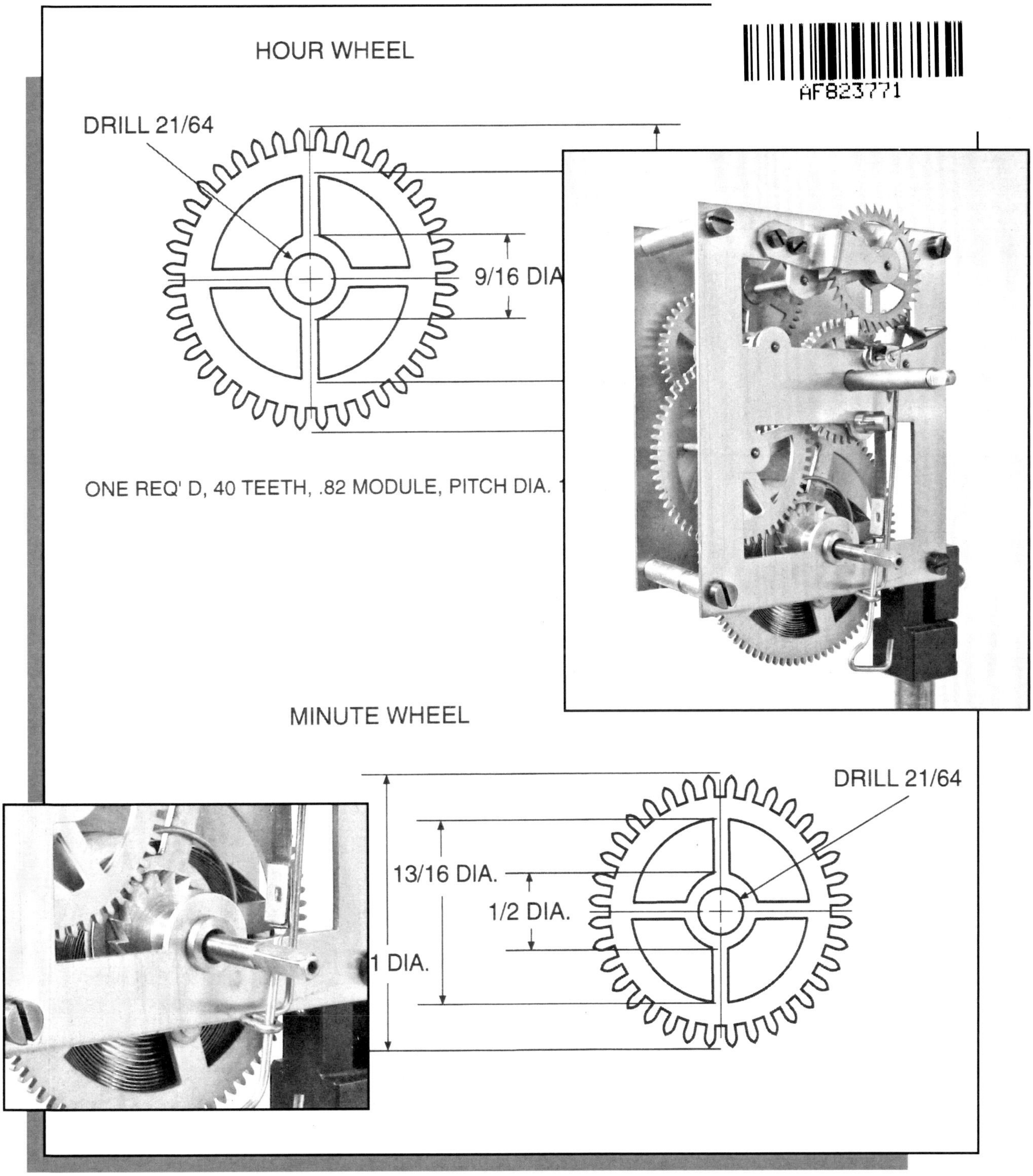

STEVEN G. CONOVER

Books by Steven G. Conover

Repairing French Pendulum Clocks $24.95
How to Repair 20 American Clocks $29.95
Clock Repair Basics $22.95
Clock Repair Skills $24.95
Striking Clock Repair Guide $22.95
Chime Clock Repair $28.50
How to Repair Herschede Tubular Bell Clocks $28.50
Building an American Clock Movement $21.95
How to Make a Foliot Clock $22.95

Clockmakers Newsletter Workshop Series:
Book 1 Repairs $37.95
Book 2 Tools, Tips & Projects $37.95
Book 3 Escapements $34.95
Book 4 Grandfather Clocks $34.95
Book 5 Tubular Bell Clocks $32.95
Book 6 Seth Thomas $37.95
Book 7 Ansonia, Gilbert & Waterbury $37.95
Book 8 Ingraham, New Haven & Sessions $37.95
Book 9 German Clocks $37.95
Book 10 Electric Clocks (to be published 2020)

New books in the *Clockmakers Newsletter Workshop Series* will be announced as they are published.

Steven G. Conover's books are available through our web store: www.clockmakersnewsletter.com.
Also available from clock parts suppliers and Amazon.com.

BUILDING AN AMERICAN CLOCK MOVEMENT

Steven G. Conover

CLOCKMAKERS NEWSLETTER

Dedicated to my wife, Karen

Published by Clockmakers Newsletter
Reading PA

ISBN 978-0-9624766-3-1

CONTENTS

The American clock movement

Description

The American clock movement featured in this book is 3-3/8" wide by 4" high. It is a timepiece movement, which means that it does not strike. The pendulum is just over 6" in length, ideal for a shelf clock. Its gear train consists of the main, second, third, fourth, and escape wheels. There are four lantern pinions in the gear train. The dial train, which enables the hands to indicate the time, consists of two wheels and two brass pinions.

Movement Specfications

Type: timepiece with recoil escapement
Module of wheels: .82
Pendulum length: 6.189 inches
Beats per hour: 9072. Calculation: (32 x 42 x 42 x 36) divided by (8 x 8 x 7) = 4536 multiplied by 2 = 9072 beats per hour, or 151.2 beats per minute.

Tooth Counts

Train Wheels

main........................84
2nd.........................60
3rd.........................42
4th.........................42
escape.....................36
center pinion............32*

Lantern Pinions

2nd arbor................8
3rd arbor.................8
4th arbor.................8
escape.....................7

Motion Work

hour wheel....................................40
minute wheel.................................36
minute pinion................................10
cannon pinion................................12

Ratios 36:12 = 3:1 and 40:10 = 4:1, 12:1 overall

* The center pinion is treated as a wheel, and it figures in the tooth count for pendulum beat (beats per hour). In contrast, the second wheel merely transmits power and is not part of the count.

INTRODUCTION

This book describes the construction of a classic American clock movement. Based on a Waterbury mechanism, the project has much in common with the millions of American clocks made between 1850 and 1930, during the clock industry's great era of mass production. Low in cost, dependable and long-wearing, these mechanisms worked for generations, and they still work today. There is no better choice for a first clockmaking project.

There is also no better way for a clock repairer to become expert at making parts for the old clocks than by making a complete movement. As each part is made and fitted into the movement, the repairer gains greater knowledge and experience that is carried over into the repair of old clocks.

Use the book by following each of the steps to finish up with a complete, working movement. As you go along, you may decide to depart from some of the individual steps, if you already have your own method. The instructions can also be viewed as a group of recipes for cutting gears, making new pinions, replacing pallets, and so on. Whenever the information is needed during the repair of an old clock, it may be possible to change a few dimensions and then proceed to make the required part. Practice the basics of lathe work, milling, and fly cutting as shown in the various steps, and you will acquire valuable skills for repair work or further clockmaking.

To make the movement, you will need at least a small lathe with milling capability, a drill press, dial calipers, and hand tools used in clock repairing. In addition, you will need several special tools and fixtures for wheel cutting and the depthing of wheels and pinions, which can be made in your shop or purchased from listed sources. You will need clockmaking brass sheet, brass rod, and steel rod, which are also available from sources listed in the book.

The lathe deserves special mention. Although many clock repairers have a watchmaker's lathe for polishing of pivots and other basic tasks, it can only be used to the fullest if the owner has several expensive accessories, including a cross slide and a milling attachment. A better choice for the clock repairer is the so-called "miniature lathe". One of these is the Sherline 3-inch lathe. Another is the Unimat-3: it was discontinued in 1990, but thousands are in use today and accessories are still available. Properly equipped, the Sherline and Unimat-3 (or other Unimat lathes) will handle the project. I use both lathes in my shop, and I do prefer the Sherline overall; but the Unimat-3 handles certain tasks better for me, such as drilling lantern pinion bodies, fly cutting ratchet wheels, and drilling with the vertical attachment.

Acknowledgements

The technical content of this book owes much to Ron Pfleger, who spent many hours helping me develop procedures and solve problems. The project would not have been completed without his help.

Our clockmaking sources deserve credit, too, for we have learned much from Laurie Penman, W.R. Smith, John Wilding and others about the various steps toward building a clock movement.

Special thanks are due to Archie Perkins for allowing the reproduction of his directions for making fly cutters; to J.M. Huckabee for permitting me to print his drawings for the wheel cutting fixture; and to the American Watchmakers-Clockmakers Institute (AWCI) who originally published their work.

Reading, Pennsylvania
Steven G. Conover, 1994

1
TOOLS, MATERIALS & SOURCES

It is frustrating to become interested in an activity such as clockmaking, only to find out that it requires one or more pieces of equipment that you don't own and will not be able to buy. That's why this American clock project utilizes the kind of tooling which is already used for clock repairing in many shops. Whenever possible, handmade fixtures and cutters are described, instead of expensive purchased items. If all these items were obtained from suppliers, the project would be beyond the budget of many clockmakers. Cost considerations aside, making one's own tools can be satisfying.

The Sherline is suggested as the lathe or lathe/mill because it works well and is currently available. The Unimat-3 is also shown because so many are still in service. Other lathes or milling machines would undoubtedly work well for the project.

The Huckabee Wheel making Fixture was designed for the American Watchmakers-Clockmakers Institute (AWCI) by J.M. Huckabee of Texas. By making this fixture, you can solve the problem of obtaining a dividing head or other indexing device, which is required for gear cutting. The plans for this fixture are found in Appendix A.

Fly cutters as described by Archie Perkins, writing in AWCI's *Horological Times*, worked very well on this project. After trying several cutter designs, I found that Mr. Perkins' cutters did not rub or hammer at the gear blanks as did other cutters. In addition to the material on cutters in Chapter 4, you will find Mr. Perkins' own directions for making them in Appendix B.

The flat depthing tool is a clockmaking rather than a clock repairing tool. The wheel and pinion are placed directly on the tool, before they are mounted to clock arbors. W.R. Smith and John Wilding have published different plans for a flat depthing tool. I found these plans very helpful in constructing the movement. References are given in this chapter.

The basic idea behind the movement was to make the subassemblies and then fit them together. To permit fitting the movement together to try the parts, I have adapted the procedures of W.R. Smith. A reamer is made from the same music wire used for the arbor. This reamer, when used to finish the hole in a wheel collet or lantern pinion, gives the part a sliding fit on the arbor. So, without riveting or staking parts, many of the subassemblies can be put together and tried. You can put the entire movement together, wind the spring just two turns of the key, and run it for hours—all without permanently attaching the pinions to their arbors! Final adjustments, such as the exact positioning of a lantern pinion on its arbor, can be made with the movement fully assembled. Then, after checking the movement for interference between parts, you can disassemble it for the fastening of various pieces with Loctite adhesive.

Leaded clockmaking brass is suggested for the wheels and plates, because it cuts so well compared to the commonly available hard yellow brass. Clockmaking brass, which is still used by the modern clock manufacturers, causes less saw blade breakage and produces a smaller burr during the fly cutting of gear teeth.

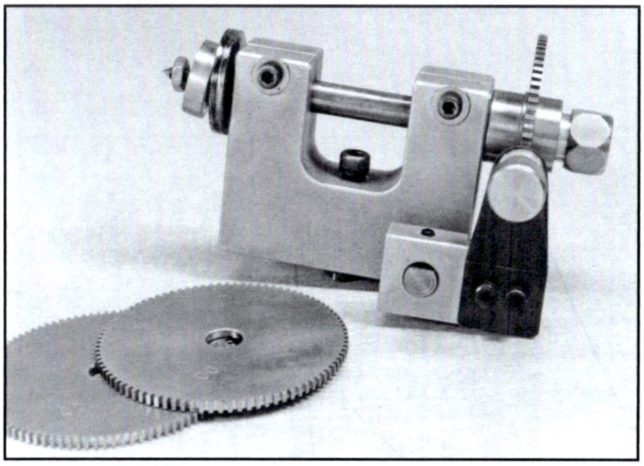

Fig. 1. The Huckabee Wheel Making Fixture.

SUMMARY OF MATERIALS

Raw Materials (Refer to Material Sources below)

Clockmaking brass, alloy C-353. Two pieces, each 4" x 12" x .062" thick	5
Brass sheet 4" x 4" x .032" thick, for escape wheel	3 5 6
Brass rod 1/4", 3/8" and 7/16" diameters, one 12" length of each	1 3 6
Drill rod 1/4" and 3/16" diameters, one 12" length of each	1 5
Music wire .040" and 3/32" diameters, one 36" length of each	1 3 6

Purchased Parts

Mainspring, loop end type, 3/4" x .0175" x 96" (alternate mainspring .018" thick)	4 5
Pallet unit (also called a verge), bent strip recoil type	4 5
Screws for pillars, steel, slotted fillister head, #4-40 x 1/4", 8 required	1 5
Screws for escape cock, steel, slotted fillister head, #4-40 x 1/8", 2 req'd (or shorten longer screws)	1 5
Screw for pallet cock, steel, slotted fillister head, #3-48 x 1/8", 1 required	1 5
Key, #6 single end, steel or brass	4 5
Hands to suit your dial	4 5
Hand washer and taper pin	4 5
Clock oil and grease	4 5

Key to Material Sources

1. MSC Industrial Supply, 1-800-645-7270, www.mscdirect.com
2. Blue Ridge Machinery, 1-800-872-6500, www.blueridgemachinery.com
3. Your local hobby shop
4. Merritt's (clock parts) Douglassville PA, 610-689-9541, 1-800-345-4101 (order line) www/merritts.com
5. Timesavers (clock parts) Scottsdale AZ, 1-800-552-1520, www.timesavers.com
6. K&S Precision Metals (ask for a list of retailers in your area) 773-586-8503, www.ksmetals.com

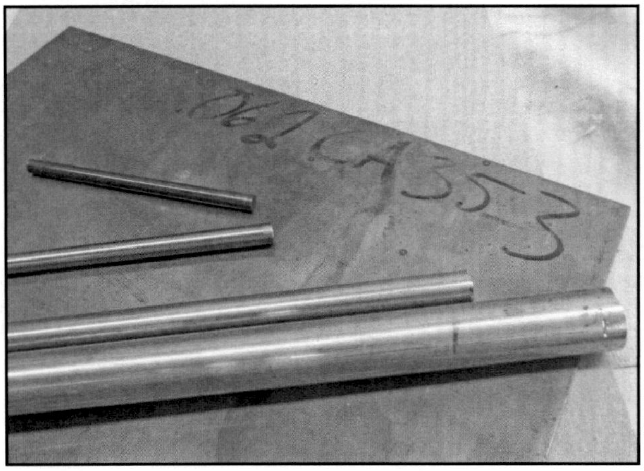

Fig. 2. Metals of the type used in the project

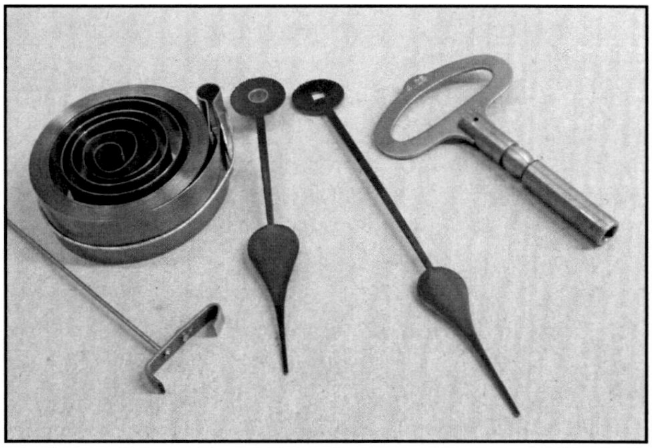

Fig. 3. Some of the purchased items

TOOLS

Tool Items

Several small lathes could be used to make parts of the movement. Here are two suggestions, including a representative selection of attachments.

Unimat-3 Lathe and Attachments*

Unimat-3 lathe
3-jaw chuck
4-jaw independent chuck
drill chuck
live center
vertical milling and drilling attachment
machine vise
indexing and dividing attachment
collet adapter and drawbar
division plate

Sherline Lathe and Attachments

Sherline items available from Tool Sources 2, 3, 4 and Sherline distributors nationwide

Sherline lathe, 4000 Series
3-jaw chuck
4-jaw independent chuck
drill chuck
cut-off toolholder
slitting saw holder
WW collet adapter and drawbar
vertical milling column

Alternate to Sherline lathe with milling column:
Sherline milling machine, 5000 Series

Other Tools

drill press..4
dial calipers showing .001" increments..............1 2
jeweler's saws and small files.................................5
Huckabee Wheel Making Fixture............Appendix A
flat depthing tool...............see Figure 4 on this page
Draucker Lantern Pinion Wire Shear......................6
Other tools for clock repair or clockmaking............5

** The Unimat-3 was discontinued in 1990. Used machines can be found, and Unimat attachments are still available, new and used.*

Key to Tool Sources

1. MSC Industrial Supply, 1-800-645-7270, www.mscdirect.com
2. Blue Ridge Machinery, 1-800-872-6500, www.blueridgemachinery.com
3. Sherline Products, Vista CA, 1-800-541-0735, www.sherline.com
4. Harbor Freight Tools, nationwide, 1-800-423-2567, www.harborfreight.com
5. clock parts suppliers: see Material Sources on page 2
6. No commercial source is known for the Draucker pinion wire shear. However, Wim Vander-Reyden created plans for making a similar tool: see pages 58-60 of Steven Conover's *Book 2 Tools, Tips & Projects* from his *Clockmakers Newsletter Workshop Series*.

Fig. 4. Flat depthing tool (John Wilding design). Plans can be found in Mr. Wilding's book, The Construction of an Elegant Scroll Type Skeleton Clock. *Laurie Penman's book,* Making Clocks, *includes plans for making a similar flat depthing tool.*

Fig. 5. Machining with a Sherline lathe.

2
PLATES & PILLARS

Leaded clockmaking brass, 1/16" (.062") thick, is suggested for the plates, since it can be drilled, cut, and filed so much more easily than common sheet brass. Saw out two pieces of brass to the size of the plates, leaving a small amount of extra material all around to allow for finishing. Fasten the plates together with the clamps which will hold them during the drilling of the pillar holes. File the edges of the plates, checking for square, straight sides. Now coat one of the pieces with layout dye on one side. With a marker, write "top front" and "top rear" on the plates. From this point onward, the plates will go together only one way. Using a steel rule and a sharp scriber, measure and draw in the centers for the four pillar holes. Scribe the lines onto the front plate, which has been marked with the dye.

Center punch the four hole locations. To make your own punch to be used throughout the project, refer to Figure 6. Grind off a two-inch length of the 3/16" diameter music wire and grind a point on the end. The angle should be about 60°. In the lathe, file the angle until the point is sharp and clean.

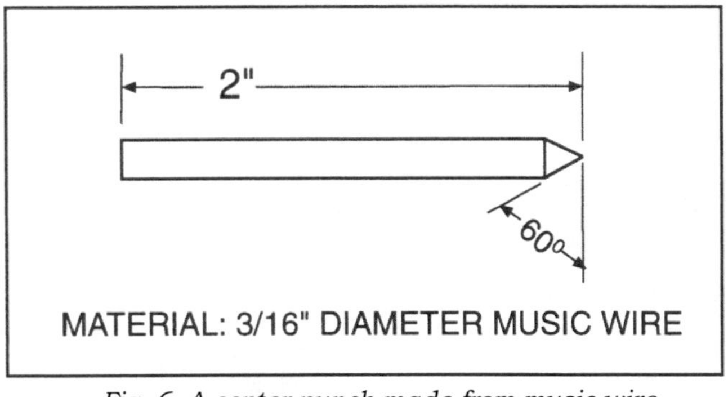

Fig. 6. A center punch made from music wire.

In order to place the center punch accurately on the crossed lines which represent a hole location, use a 3-power loupe. Tap the punch lightly with a hammer, then check to see whether the mark is exactly at the crossing of the two lines. If it is not, punch at a slight angle to "move" the mark to the right place. Then strike sharply to make the mark deeper. A machinist or toolmaker would use two punches to mark a hole; first a prick punch to locate the hole, then a center punch (with a different angle on the tip) to make the mark deeper. However, the handmade 60° punch works well on our project, and it is used later to accurately mark the pivot hole locations.

Drilling the Pillar Holes

With the plates still clamped firmly together, go to the drill press and set up to drill the first pillar hole. It does not matter which hole will be done first. It is important to support and clamp the work to avoid having a drill grab in the work and stall the drill press. The brass piece can even be swung out to the side, where it can cause severe injury.

With this in mind, it is a good idea to modify the drill to make it less likely to grab in the brass. The modification will make the drill much safer to use, but it will no longer be any good for drilling steel. As shown in Figure 7, place a drill flat on the bench, and rotate it until one cutting edge, at the tip, is parallel to the bench top. Make a few light strokes with a sharpening stone across the lip, not on the spiral flute of the drill. Hold the stone parallel to the bench top as you bring it across the steel. Rotate the drill and repeat the process on the other lip of the drill. Small drills, less than about 1/8", may not need this treatment, but larger drills certainly benefit from it.

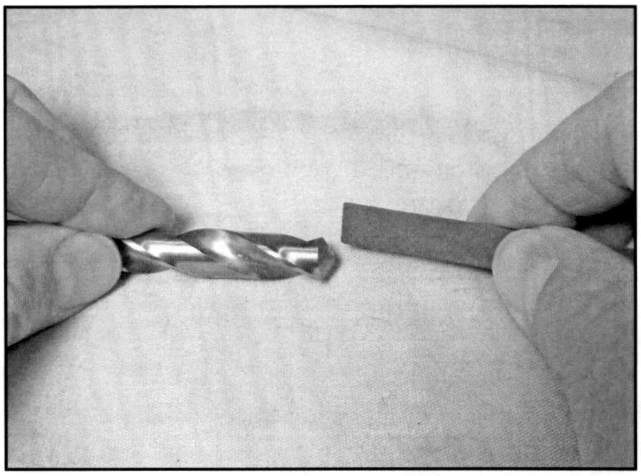

Fig. 7. Preparing a drill for brass.

The plates should be clamped to each other, as mentioned. You may have your own way of working

FIG. 8. PLATES & LOCATING PINS

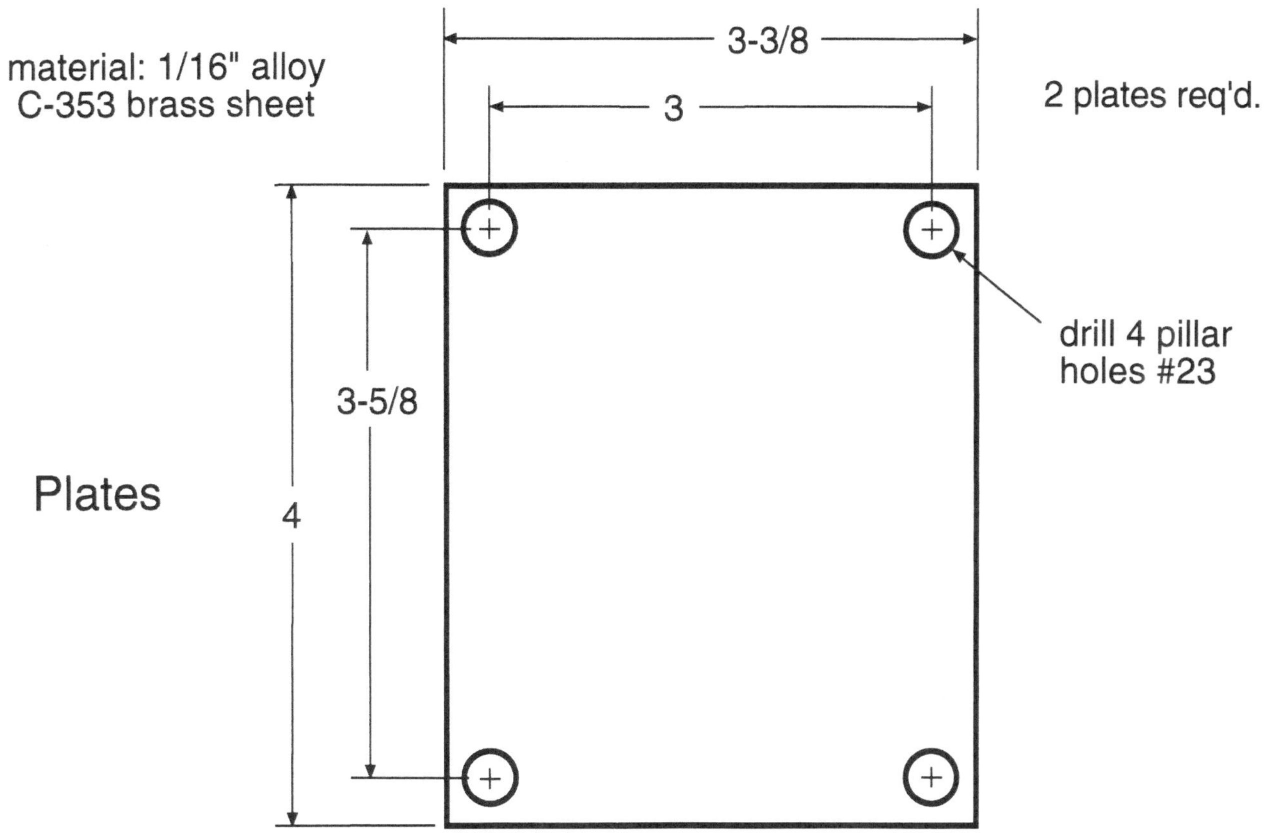

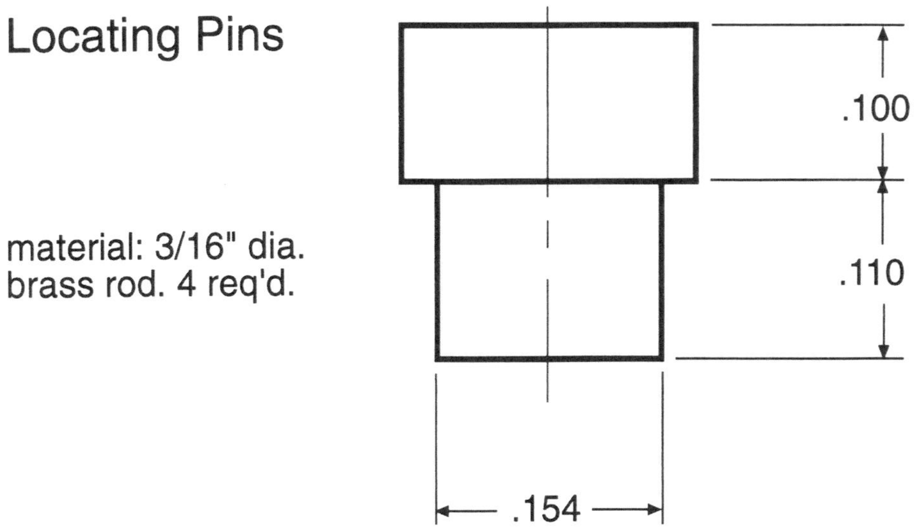

FIG. 9. PILLARS

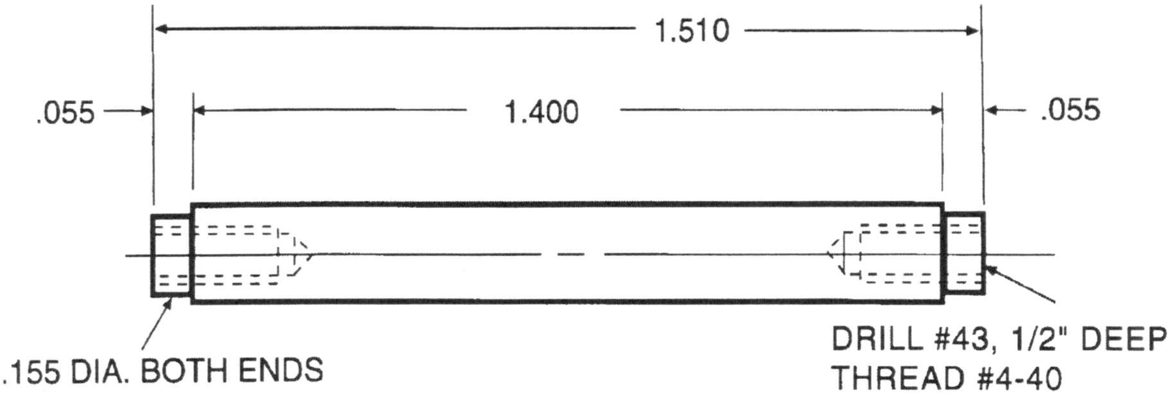

TOP PILLARS (2)
MATERIAL: 1/4" DIA. BRASS ROD

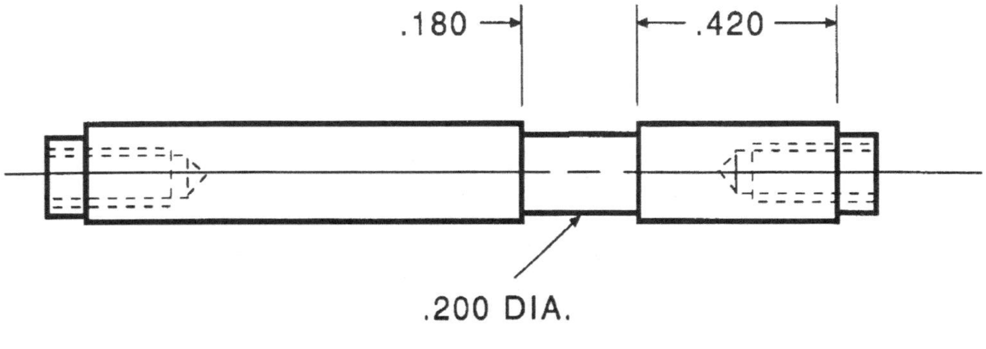

BOTTOM PILLARS (2)
SAME DIMENSIONS AS TOP PILLARS,
EXCEPT FOR .200 RECESS.
MAKE 1 FROM 1/4" DIA. BRASS ROD,
1 FROM 1/4" DIA. DRILL ROD.

NOTE: THE STEEL PILLAR MAY BE TOO LARGE IN DIAMETER TO ACCEPT THE LOOP-ENDS OF SOME MAINSPRINGS. IF NECESSARY, TURN DOWN THE DIAMETER OF THE STEEL PILLAR FROM 1/4" TO .235"—.240" TO FIT THE LOOP END.

with the drill press, but you should consider also clamping the plates to the drill press table. This will help assure that the work will not swing, if the drill catches in the brass.

Begin by installing a center drill in the drill press chuck and spotting the first hole at the punch mark. Now drill the hole with a #23 drill which has been prepared for brass work as described above. Remove the plates from the drill press.

Using The Locating Pins

Figure 8 gives the dimensions of the pins, which are made on the miniature lathe. A locating pin should fit so that a very light tap with a small hammer will seat it. Install a locating pin in the hole just drilled, through both plates. Keep the plates clamped together and drill another of the #23 pillar holes. Install a locating pin in this hole. The locating pins assure that the pillar holes will all line up when the drilling is finished. Proceed to drill the other two holes, installing a locating pin each time.

The plates are now ready for the pillars. With the pillar holes drilled and the plates marked to identify the top front and top rear, they can be taken apart and reassembled the same way each time. Later in the project, the pivot holes and other threaded holes will be made. In addition, the front plate will have sections cut out to make the wheels more visible.

Making the Pillars

There are four pillars for the movement. The bottom pillars have a recess cut in them to allow the main wheel to pass. Three of the pillars are brass. The fourth, which is on the bottom left when the movement is viewed from the front, is made of drill rod (steel). All the pillars could be made from drill rod, but three were made from brass to make them easier to machine. The fourth pillar is made of drill rod to safely hold the mainspring loop end.

To make the pillars, install a length of 1/4" rod in the 4-jaw chuck and indicate the piece. (The 3-jaw chuck will not make concentric pillars, and the plates will not line up.) Turn down one end to .155" diameter for a slip fit in the pillar hole in the plate. Using a #43 drill as the tap drill, drill the end for the #4-40 threads. Tap the hole now if the chuck is strong enough to hold the pillar; otherwise, tap the threads later. Turn the piece around in the chuck, indicate, then part off the excess material so that the entire piece is 1.510" long. Make the other turned-down end, drill it, and tap the #4-40 threads. The bottom pillars need to have a recess cut into them, which can be made easily in the lathe by using a parting tool. It is important that the pillars have the same length of 1.400" between shoulders to allow the plates to fit properly.

Assembling the Plates and Pillars

After the pillars are completed, assemble the plates and pillars. Use ordinary screws with either slotted or socket heads; the plates will be taken apart and reassembled many times during construction, and there is no reason to risk scratching handmade screws at this stage. Use a washer under each of the eight pillar screws. The final screws, if hand made, will not require washers.

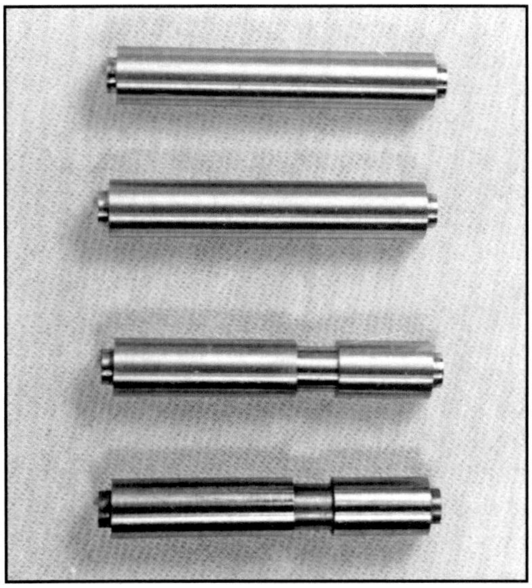

Fig. 10. The four pillars.

Fig. 11. Plates and pillars assembled together.

3
WHEEL CUTTING WITH THE HUCKABEE FIXTURE

There are several types of lathe or mill accessories which will produce clock wheels. The one which is described in this chapter is the Huckabee Wheel Making Fixture (Figure 12), designed for AWCI by J.M. Huckabee of Texas. It was selected because it can be made, rather than purchased, by the clockmaker, and because it works so well. Complete plans for the Huckabee Wheel Making Fixture are presented in Appendix A.

The purpose of the wheel making fixture is to hold a wheel blank steady while a tooth (actually, the space between two teeth) is cut. Then an index pin

Fig. 12. The Huckabee Wheel Making Fixture.

is pulled free of an index plate at the other end of the fixture's spindle, which is then advanced the distance of one tooth. The next tooth space is cut, and so on. The clockmaker makes his or her own index plates with the necessary tooth counts, using an old wheel with the correct count as a pattern.

The Huckabee fixture was designed for miniature lathes, and it will fit the Sherline lathe with vertical column, Sherline mill, or Unimat-3 lathe equally well. Different hold-down bolts (also called T-nuts) are required for Sherline and Unimat. The fixture enables the clockmaker to perform indexing to cut wheels, without purchasing expensive lathe or mill attachments.

This chapter presents a wheel cutting procedure which can be followed in making the American clock movement. If you already have a wheel cutting setup, then skip to Chapter 4, which contains the specifics of the American clock wheels. Otherwise, it is time to work out the equipment and procedures to be used in cutting the wheels.

Making the Wheel Blank

To begin making a wheel blank, cover a section of the brass with layout dye. Punch mark the wheel center and use dividers to scribe the wheel blank diameter. To create a safety margin for sawing out the blank, scribe the circle 5-10mm larger than the finished blank will be. Drill the center hole with a drill

Fig. 13. Drilling the wheel hole in a drill press.

press, as shown in Figure 13. The drawings for each wheel in the movement state which size drill to use for the center hole. Observe all safety precautions for the drill press, including the modification of the drill for use on brass, as described on page 4. You may want to cut two wheels at a time. The extra becomes a spare and insurance against spoiled wheels. It re-

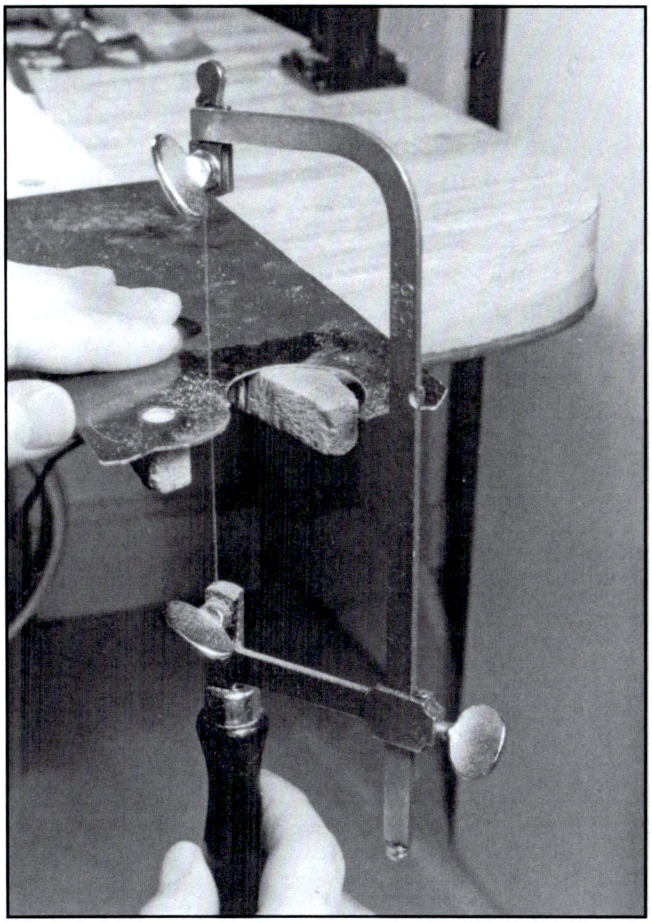

Fig. 14. Cutting out the wheel blank with a jeweler's saw.

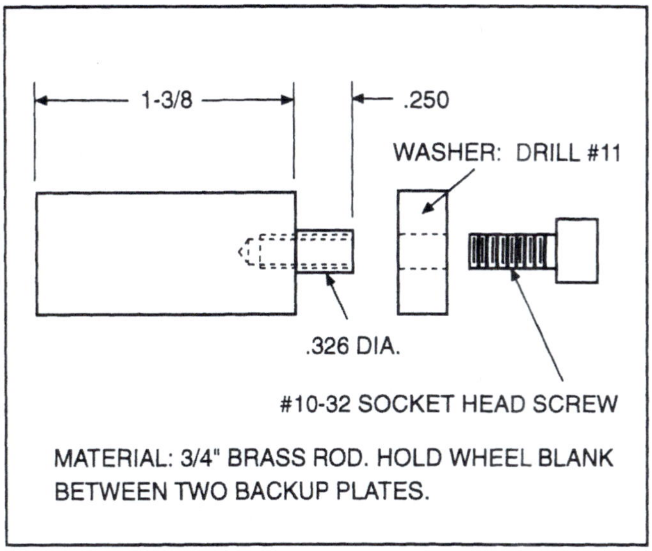

Fig. 15. A holder for wheel blanks.

Fig. 16. The wheel blank holder drawn in Figure 15, including a wheel blank and two backup plates.

quires more work to make multiple wheels at once, but it takes far less effort than redoing the complete setup to duplicate a wheel.

Now cut the wheel blank(s) out of the brass sheet with a jeweler's saw, as shown in Figure 14. At this point you will appreciate the ease of sawing the blank from clockmaking brass. Cut as close as possible to the scribed circle. Try to avoid cutting inside the line, but the few extra millimeters included in the marked diameter will allow some leeway.

The next step is to turn the wheel blank to the exact diameter from which the teeth will be cut. This can be done in the lathe, using a small wheel blank holder which is easy to make. See Figures 15 and 16. The holder consists of a piece of 3/4" diameter brass rod, two backup plates, a brass washer, and a steel holding screw. The rod is turned down at the end so the wheel blank will bc a slip fit. The end is drilled and tapped for the holding screw. In use, the wheel blank is supported between backup plates, then held with the washer and screw. It is essential that the hole in the wheel blank is concentric with the O.D. of the blank, so the brass body of the holder is placed in the 4-jaw chuck. A dial indicator is set on the .326" diameter portion, and the chuck is adjusted to run true.

Figure 17 shows two blanks mounted together on the fixture. Run the lathe at low speed, which is required by the relatively large size of the blanks compared to the usual rod stock used in clockmaking, such as 1/4" or 1/2" diameter. Take light cuts of about .005" as indicated on the handwheel, which will take .010" off the diameter. A number of passes will be required to reach the desired diameter. Check carefully with dial calipers which read in thousandths of an inch, as shown in Figure 18. When the final diameter is reached, the blank is complete. A diameter of .001" over or under is acceptable.

If the wheel blank had a center hole drilled #7, it would fit on the spindle of the wheel making fixture, which is specified as .199". Obviously, we cannot be limited to making wheels with a center hole of just

Fig. 17. Wheel blank being cut to size on the Unimat-3.

Fig. 18. Checking the diameter of the wheel blank.

one size. Mr. Huckabee, who designed the fixture, enlarges the center hole after the teeth have been cut. In one of his AWCI videos, he mounts the finished wheel concentric with the center of the lathe, and then he bores the hole.

I used another method for sizing the wheel hole. When the brass sheet was first drilled, the hole was sized for the exact diameter required for the wheel collet. The hole cannot be made smaller than a slip fit for the end of the wheel making fixture, which means the smallest wheel hole we can use is #7 (.199"). The hole can be larger, however, for example 5/16". It is only necessary to make a brass bushing for the wheel blank. The O.D. of the bushing is a slip fit in the wheel hole; the I.D. is a slip fit on the spindle of the wheel making fixture. A few bushings will permit several sizes of wheel hole.

Setting Up the Fixture

Whether or not a bushing has been used, the next step is to mount the blank on the wheel making fixture, as shown in Figure 19. Coat the edge of the blank with layout dye. Two backup plates are usually needed, each drilled to be a slip fit on the spindle. The wheel blank is held between these plates, which provide rigidity for the setup. A thick brass washer is placed on the spindle, followed by the nut. For maximum rigidity, the diameter of the backup plates should be just small enough so they will not be touched by the fly cutter as the teeth are being cut.

However, you may be cutting two wheel blanks, each 1/16" thick, at the same time. This makes it less important to have backup plates that are as

Fig. 19. A dial indicator checks the blank.

large in diameter as possible, for the blanks themselves are rigid enough. Sometimes it will be possible to use only one backup plate if the blanks are going to fit on the spindle of the wheel making fixture. If you have made a bushing to accommodate a wheel hole larger than. 199", you must first put a backup plate with a .199" diameter hole onto the spindle, to provide a seating for the wheel blank.

In Figure 19, the Huckabee fixture is mounted on the lathe. A dial indicator is touching the blank. If the fixture's index pin is pulled back, you can rotate the spindle and check the blank for runout. A thousandth or two of runout can often be corrected by loosening the blank in the fixture, rotating it, and retightening. A larger amount of runout means that the blank was not firmly supported as it was turned down to final diameter. Such a blank cannot be used as it is, although it could be recycled by later turning it to a smaller diameter for another wheel.

Fig. 20. The Unimat-3 set up for wheel cutting.

Figure 20 shows the fly cutter mounted in a holder. The Unimat-3 is shown, with vertical column and drill chuck, gripping the holder. The fly cutter and cutter holder will be described in the next chapter.

A T-nut attaches the wheel making fixture to the cross slide. A small square can be used to check that the fixture is lined up with the cross slide and lathe bed. Check the placement of the fixture to make sure you haven't set it too far in or out to accommodate the index plate and the wheel blank. It's a good idea to look over any machine setup before spending time adjusting other items that might have to be done over.

The cutter must be set on center. This is easy to accomplish, because the spindle of the wheel making fixture has a male center turned on the end. Move the cross slide with the longitudinal and cross slide handwheels until the tip of the fly cutter is lined up exactly with the center on the fixture, as shown in Figure 21. Now push down on the drill press handle of the Unimat, or turn the vertical handwheel on the Sherline, until the point of the center is lined up exactly with the center of the cutting face on the fly cutter. Get in close and use a loupe to avoid guesswork. When the cutter height is set on the Unimat-3, tighten the locking screw. The Sherline mill has a locking lever on the vertical column.

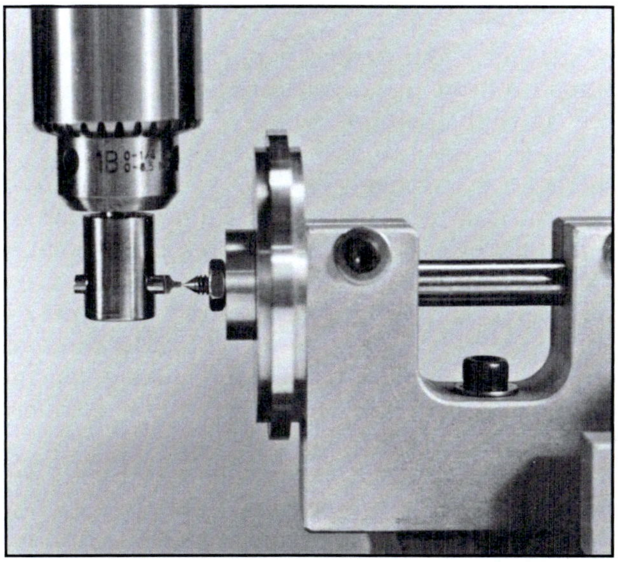

Fig. 21. Lining up the fly cutter on center.

Mount the index plate with the required number of divisions onto the wheel making fixture, tighten it securely, and place the index pin in one of the slots in the index plate, as shown in Figure 22. The lathe or mill motor is still off! Use the machine's handwheel to bring the fixture in position so the flat face of the cutter rests against the blank. At this point, get an initial estimate of the depth of cut that will be taken. Set the depth so that it appears the curve of the tooth tip will remain unformed or just started. The final depth of cut is established by trial cuts, not by measurement.

Fig. 22. Setting the index pin.

Fly Cutting

With the motor still off, use the lathe's longitudinal handwheel to move the Huckabee fixture to a safe starting position where the cutter, if rotating, would not touch the wheel blank. Put on safety glasses. Now turn the motor on at full speed, which is approximately 2000 RPM for the AC motor on my Sherline. If you are using a Unimat-3, switch the motor on at the fast setting. You are ready to start cutting. Turn the longitudinal handwheel so slowly that it may take eight or ten seconds, after contact, for the cutter to pass through the blank. The chips which are thrown off the cutter will soon accumulate

Fig. 23. Starting at the top of the photo, the tooth spaces were cut progressively deeper until, after approximately eight were done, the depth was correct.

into a covering of very fine, almost powdery flakes on the cross slide and the bench. Thicker chips and the sound of hammering indicate the cutting operation is not going properly. It may be lack of clearance on the cutter, too slow a spindle speed, or too fast a feed rate through the blank which is to blame.

My Unimat-3 motor must be stopped at the end of each pass. Turn the cutter away from the blank, then crank the unit back to the starting position. The motor is turned on for the next pass of the cutter. If the fly cutter is run back through the blank under power, the cutter will break, because of play in the setup. No such problem exists with the Sherline, which can be run as the cutter is brought back

Fig. 24. End view of the Huckabee Wheel Making Fixture, with the index pin seated in the index plate.

through the tooth space just cut, to reach the starting position. Still, the fly cutting is a slow process requiring a full minute (including indexing to the next position) to fly cut through a 3/16" blank.

The depth of cut must be carefully adjusted. After the first cut, which should be a shallow one, stop the motor and set the depth of cut a few thousandths of an inch deeper. Cut the next tooth space and again check the result. Figure 23 shows the way each tooth is cut deeper than the one before it, until the correct depth is found after six or eight teeth. The depth is correct when the faintest possible line, a few thousandths of an inch wide, is left at the tooth tip. If the teeth are brought instead to a sharp edge, with no layout dye showing at the tip, I have found that a minimum of four thousandths will be lost off the O.D. of the wheel.

After the correct depth is established, it is helpful to use a black marker to identify the first full-depth tooth space. Cut all the way around the wheel, finishing up by recutting the partial depth teeth. The marked tooth space is the one that is not necessary

Fig. 25. This group of wheels was cut with the Huckabee fixture and a lathe such as the Unimat-3.

to recut. I have found this a better method for setting the depth of cut than recutting two adjacent tooth spaces until the depth is found.

It is a good idea to have as many index plates as possible for the Huckabee Wheel Making Fixture, to reduce the number of occasions where you must use, for example, every third division on a plate to cut 40 teeth from a 120-tooth index plate. Wheel cutting may become so boring and repetitive that you can make a mistake counting to three! Figure 24 shows an end view of the fixture with the index pin seated in an index plate of 84. This plate can be used for a wheel of 84, 42, 28 or 21.

Figure 25 shows a collection of wheels cut with the Huckabee fixture. Some of the wheels are still coated with layout dye, and all are ready for crossing out the spokes and mounting on wheel collets.

4
WHEELS & CUTTERS

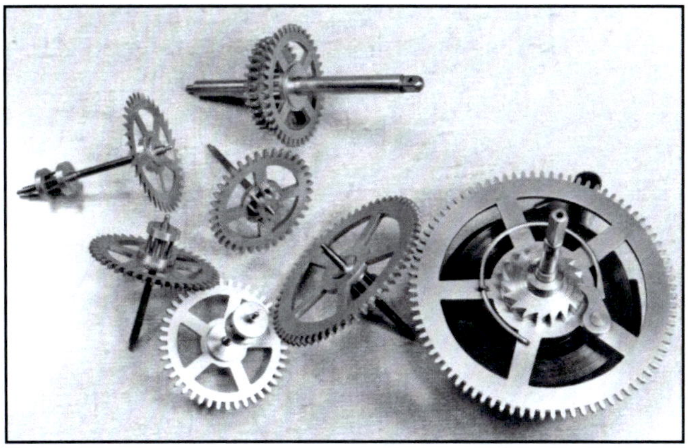

Fig. 26. The completed wheels.

There are seven wheels in the clock train: main, second, third, fourth, escape, hour and minute. Another gear, the center pinion, is cut as a wheel and treated as such in calculating the beats per hour of the train, but we will cover it in a later chapter. Figure 26 shows the seven completed clock wheels with pinions and other parts attached.

To make the construction as simple as possible, I decided to use just two center hole sizes, 21/64" and 3/8". The 21/64" size hole is used in all the wheels except for the main wheel, which is 3/8".

The wheels must be "crossed out"—that is, spokes must be cut. This gives the wheels a finished appearance. Crossing out is also necessary to reduce the mass of the wheels, which start and stop with each beat of the escapement. The higher up in the train, the more important it is to have the lightest possible wheels. The only wheel which could, if desired, be left solid is the main wheel, but it is always crossed out in American movements. Dimensions and procedures for the wheel crossing work are given in this chapter, but the actual crossing out, for all the wheels except one, can wait until near the end of construction. This not only keeps the project moving along (making the spokes is time consuming) but eliminates the possibility of dropping or otherwise damaging a wheel which has had hours of finishing work applied to it. It is suggested, therefore, that the crossing out be done just before final assembly. The exception is the main wheel, which must be crossed out as you work through the next chapter. We cut the main wheel teeth in this chapter (along with the other wheels), then cross out the main wheel and add the click rivet and click spring in the next chapter.

Gear Calculations

The metric module system is used to size the wheel teeth. Module is simply pitch diameter, which is the diameter at the circle where meshing gears touch, divided by the number of teeth. The module of our movement is .82.

Wheels such as the one shown in Figure 27 can be cut only after the blank is turned to the exact diameter first. Therefore, one of the first pieces of information we need to have for each wheel is the outer diameter, so that we can make the wheel blanks using the method discussed in Chapter 3.

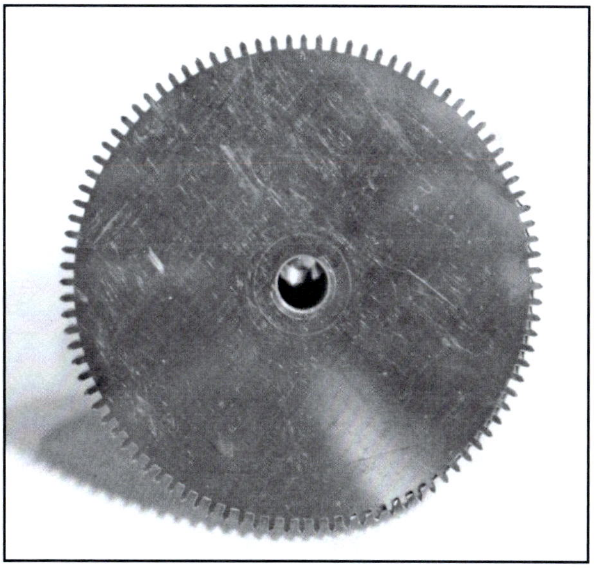

Fig 27. A clock wheel.

The method of calculation is not difficult: just take the tooth count, add a factor to it, and multiply the result by the module. The answer will be metric, which I have converted to inches for this project. As an example, the main wheel has 84 teeth. Add the factor, which is 2.76 for all the wheels, to obtain 86.76. Multiply 86.76 by the module of .82 to yield the blank diameter of 71.1432mm. Dividing this by 25.4 converts it to 2.801 inches for the main wheel. The outer diameter of the wheel blanks for the clock is given in the drawings for each wheel, Figures 35 through 38 and Figure 47.

Fly Cutter for Wheels

One fly cutter is used for all the wheels except the escape wheel. There are several designs for fly cutters which can be made to cut clock gears. The type which I found to work best was described by Archie Perkins in AWCI's magazine *Horological Times,* in the

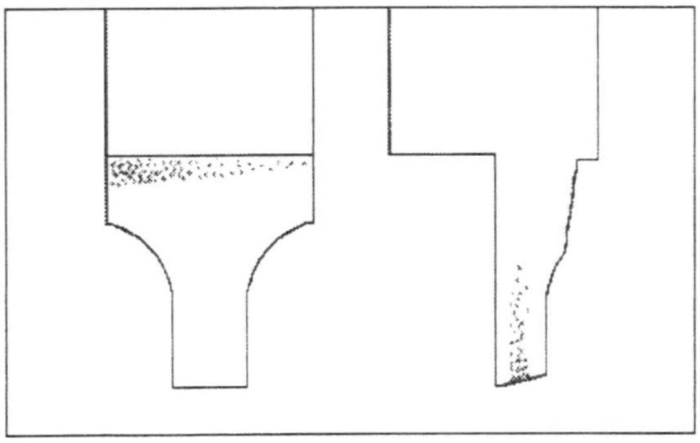

Fig. 28. The fly cutter for wheels.

August 1989 issue. An excerpt from the article, in which Mr. Perkins explains his step-by-step method for making cutters, is included as Appendix B at the end of this book.

Figure 28 shows front and side views of this type of cutter. Although Mr. Perkins' method is complete, I have included some additional comments here.

Figure 29 shows the way tooth space and tooth height are measured, and the way in which the cutter matches the tooth form by fitting in the space between two teeth. The distance between two teeth equals 1.57 times the module. Our module is .82, so

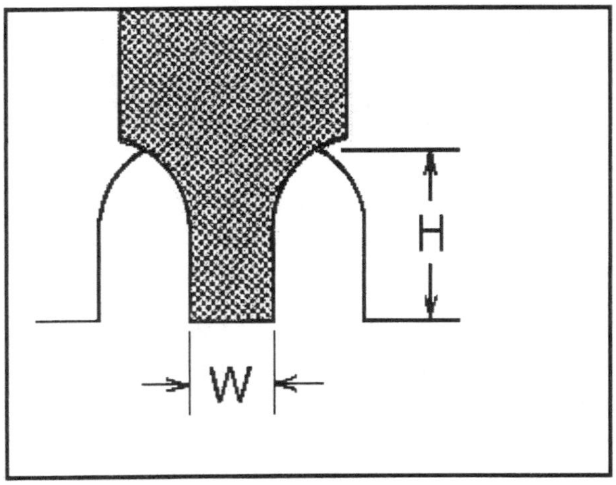

Fig. 29. Tooth space and tooth height as they relate to the shape of the cutter.

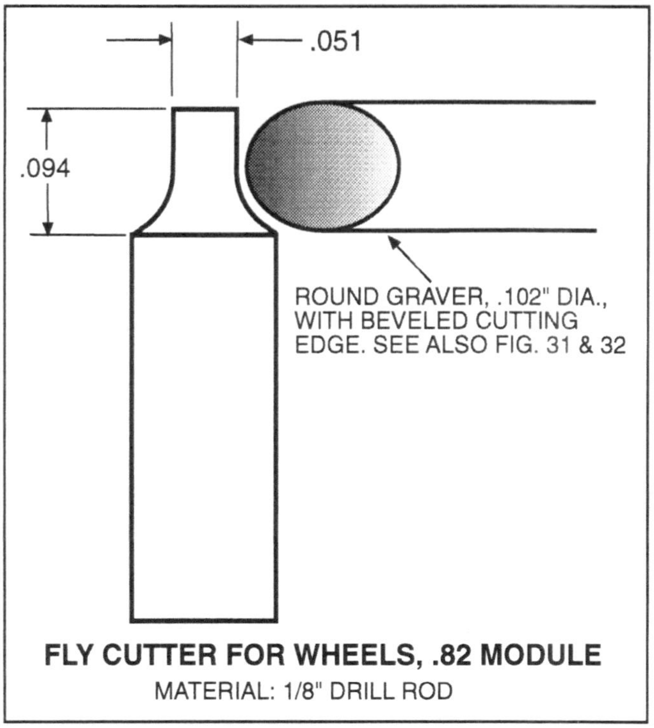

Fig. 30. Dimensions of the .82 module cutter.

the product is 1.2874mm, which converts to .051". This means the cutter is to be turned down to a width of .051" at the end. The tooth height, which is included in the finished portion of the cutter, is 2.92 times the module, or 2.3944mm, which converts to .094".

Figure 30 shows the dimensions of the .82 module fly cutter required for the project. It is made from a piece of 1/8" diameter drill rod. This same cutter will make the teeth for six of the seven wheels in the clock train. The drawing also shows how a round graver matches the curve of the tooth.

Fig. 31. Shaping the cutter.

Figure 31 shows the fly cutter being shaped in a watchmaker's lathe, although one could just as well use a miniature lathe fitted with a tip-over T-rest. If you have never made such a cutter before, it may take an hour or two. A lot of care is required to make

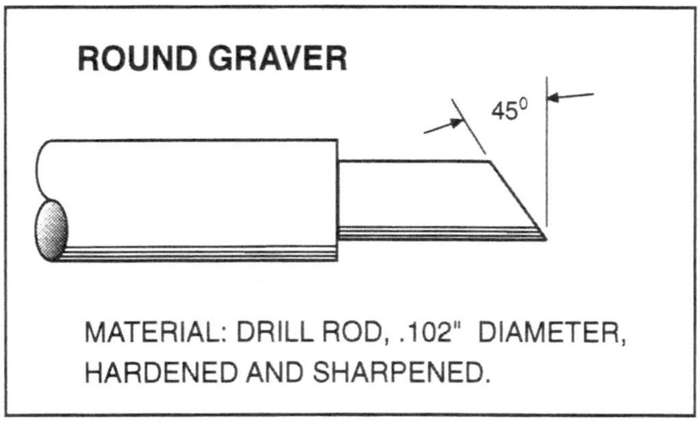

Fig. 32. Dimensions of the round graver.

the right dimensions on the cutter. Frequently, you will have to check the width of the cutter at the tip, as you approach the .051" dimension. Speed in making fly cutters comes only with the practice gained in making a number of them. The round graver, which Mr. Perkins describes in Appendix B, is being used to turn the tooth radius onto the cutter blank. This particular graver does not guarantee that the correct radius will be made on the cutter blank, which would automatically create the right tooth form. It does, how ever, make the job of turning the radius much easier, since the desired curve is already on the graver. While it is true that rotating the cutter slightly will alter the radius produced on the cutter blank, at least you are not struggling with a sharp pointed graver or a round graver that is too large.

Fig. 33. Sharpening the cutter.

The round graver to be used in making the .82 module fly cutter is drawn in Figure 32. To make this graver, take a 5" long piece of drill rod and turn the end to a diameter of .102", which is twice the radius of the tooth (also equal to twice the width of the tooth space). According to Mr. Perkins, the end of the graver is ground off at a 30° to 45° angle, hardened, and sharpened on a stone. When applied to the cutter

Fig. 34. The cutter is shown in a holder, which is mounted in a drill chuck in the Sherline mill.

blank, the round graver helps, as explained above, to produce the correct radius of the tooth form, which is .051".

After the tooth form has been placed on the cutter blank, bring it to a high polish. Then, following Mr. Perkins' directions in Appendix B, file away one half the diameter of the cutter. Add the all-important reliefs which will allow the cutter to produce gear teeth without rubbing or banging on the gear. Harden and sharpen the cutter using Appendix B as a guide. Figure 33 shows an alternate method of sharpening the cutter by stroking it with a hand held stone. Care is required to avoid rounding the edges of the cutter.

When installed in the lathe (with vertical column) or milling machine, the cutter is secured in a holder which was described in Chapter 3, "Wheel Cutting With the Huckabee Fixture". Figure 34 shows the cutter mounted in the Sherline equipment.

Figures 35 through 38 on pages 16-19 give the dimensions of six of the clock train wheels. Proceed to cut these wheels using the detailed method presented in the previous chapter.

The present chapter then continues on page 20 with instructions for making the escape wheel cutter and the escape wheel.

FIG. 35. MAIN WHEEL

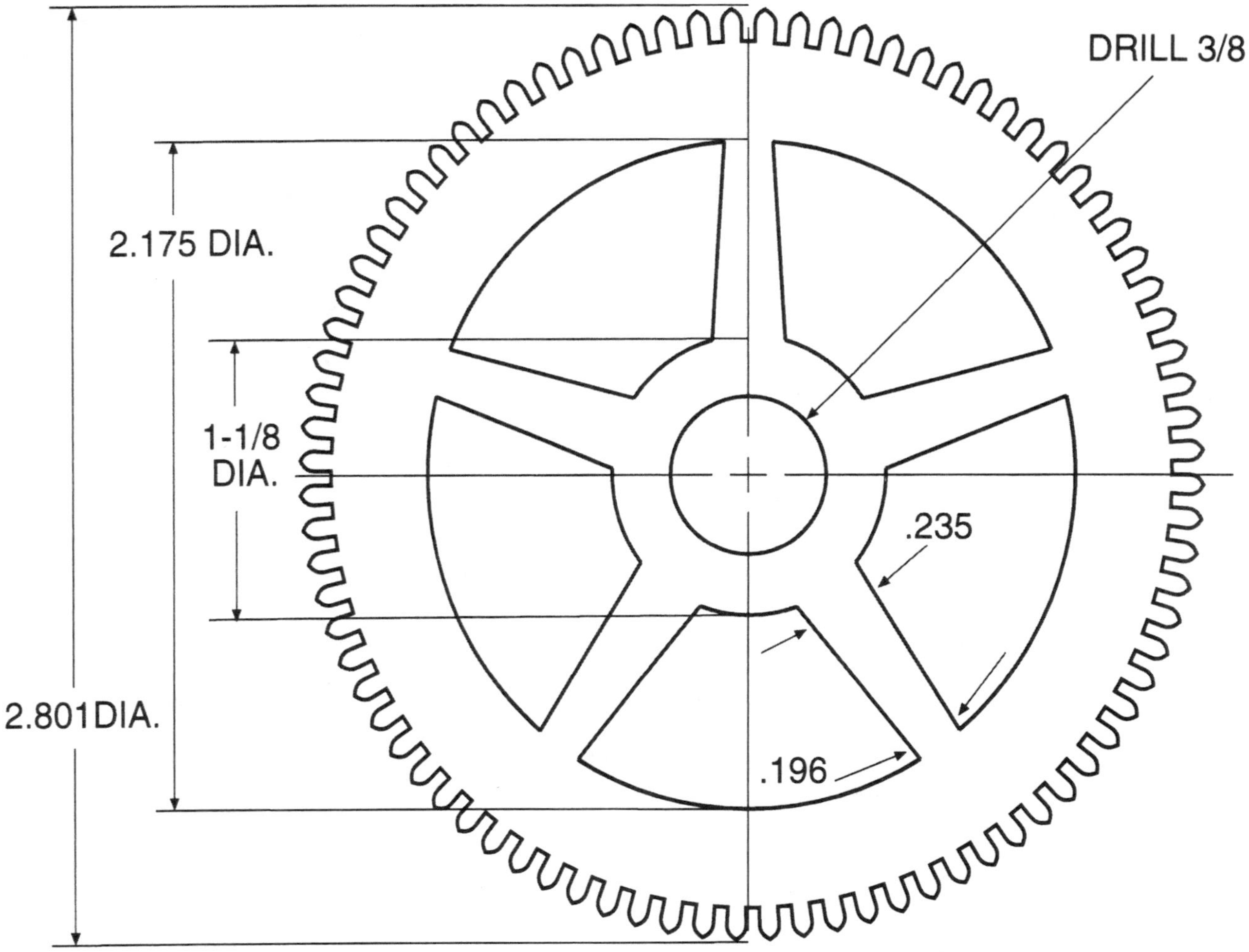

84 TEETH. MATERIAL 1/16" BRASS SHEET, C-353. ONE REQUIRED. MODULE .82. PITCH DIA. 2.711". DIMENSIONS ARE IN INCHES.

NOTE: AFTER CUTTING THE TEETH, CROSS OUT THE WHEEL IF DESIRED. AS AN ALTERNATE, WAIT UNTIL CHAPTER 5, WHEN THE OTHER PARTS OF THE MAIN WHEEL ASSEMBLY ARE ADDED.

FIG. 36. SECOND WHEEL

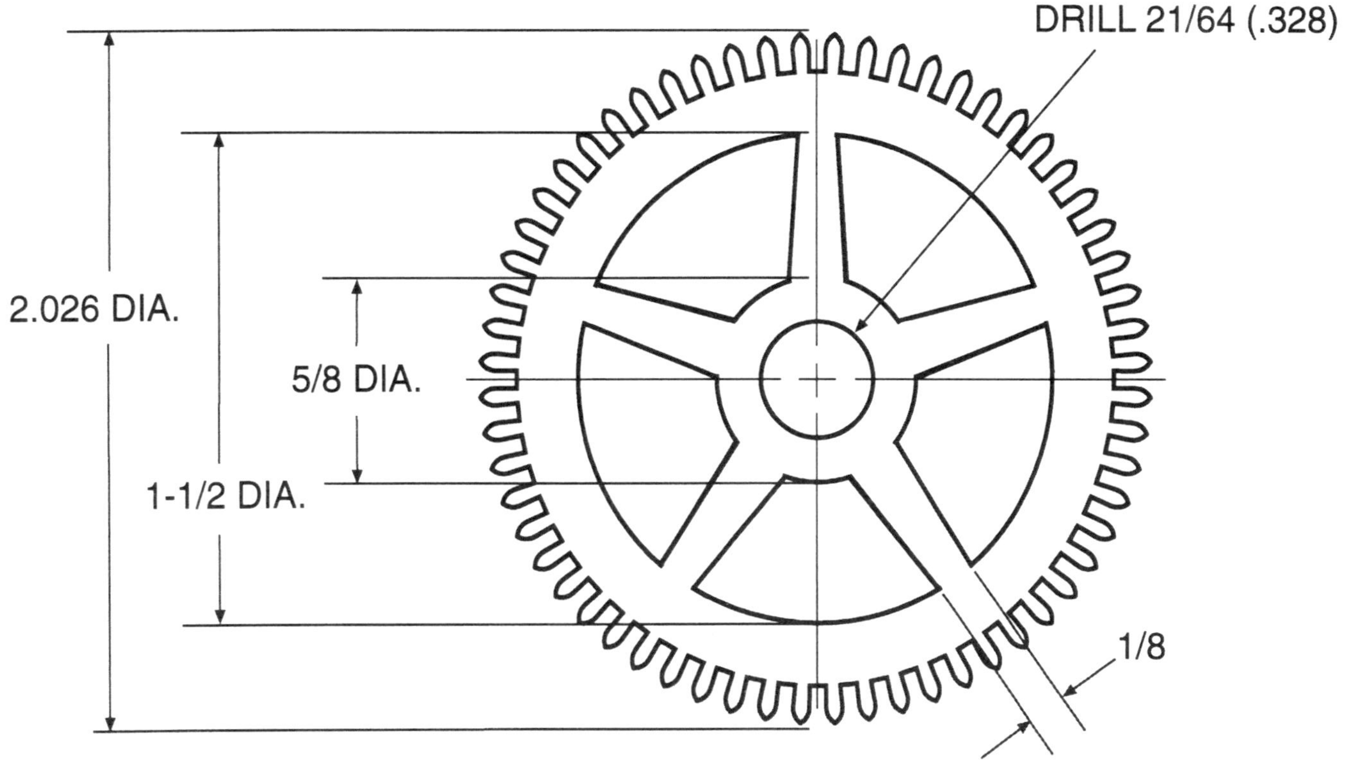

60 TEETH. MATERIAL 1/16" BRASS SHEET C-353. ONE REQUIRED. MODULE .82. PITCH DIA. 1.937". DIMENSIONS ARE IN INCHES.

NOTE: THE WHEEL CROSSINGS (SPOKES) ARE SHOWN TAPERED OUTWARD FROM 1/8" WIDE AT THE WHEEL RIM, BUT THEY CAN ALSO BE MADE 1/8" WIDE WITHOUT A TAPER. STRAIGHT CROSSINGS ARE EASIER TO LAY OUT, CUT, AND FINISH. IN GENERAL, WHEEL CROSSINGS ARE REQUIRED TO LIGHTEN THE WHEELS, BUT THE STYLE IS UP TO THE BUILDER.

FIG. 37. THIRD & FOURTH WHEELS

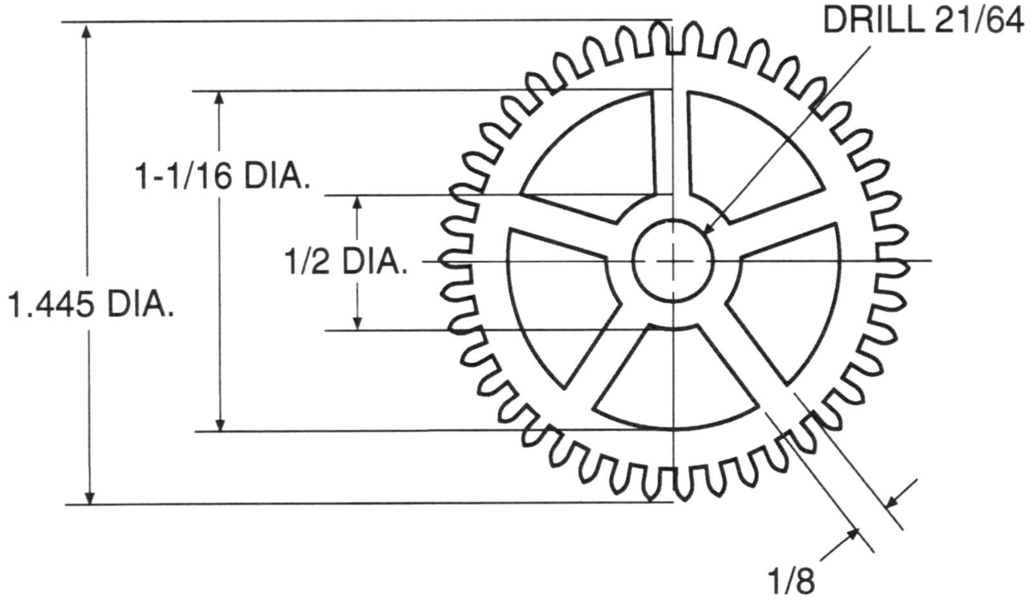

42 TEETH, .82 MODULE, PITCH DIA. 1.356". 2 WHEELS REQUIRED.
MATERIAL : 1/16" THICK BRASS SHEET, ALLOY C-353.
THE WHEELS ARE IDENTICAL AND SHOULD BE CUT AT THE SAME TIME. NOTE THAT THE SPOKES ARE SHOWN CUT STRAIGHT, NOT TAPERED, BUT THEY CAN BE TAPERED IF DESIRED.

FIG. 38. HOUR & MINUTE WHEELS

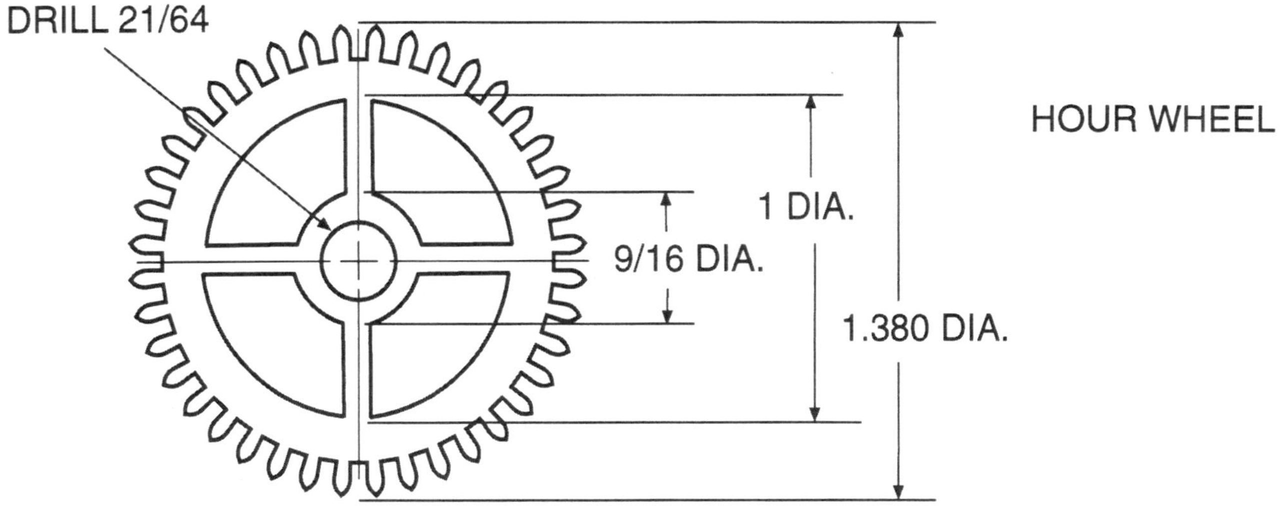

ONE REQ' D, 40 TEETH, .82 MODULE, PITCH DIA. 1.291", 1/16" BRASS SHEET, C-353

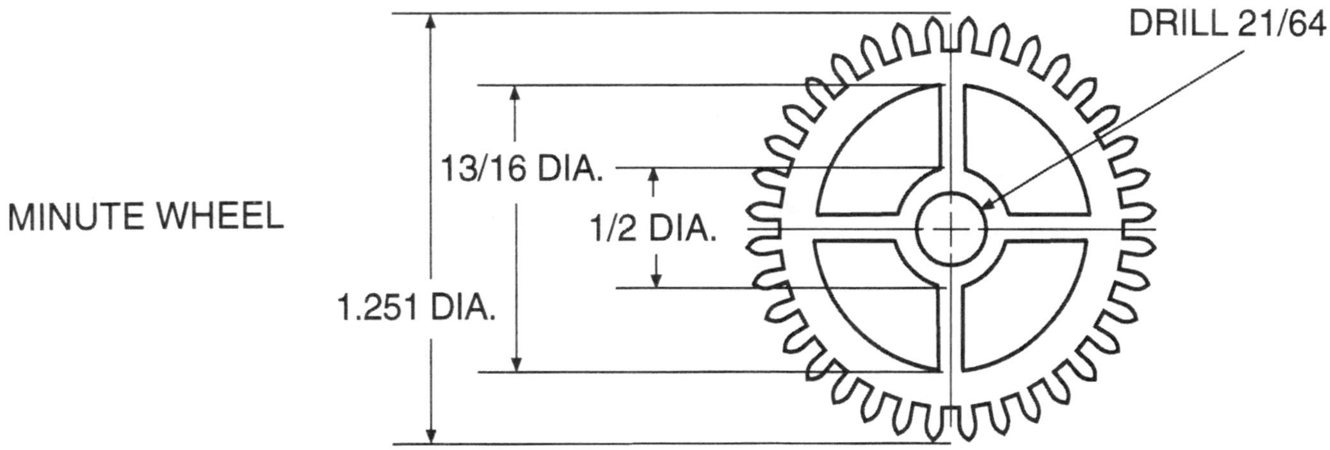

ONE REQ' D, 36 TEETH, .82 MODULE, PITCH DIA. 1.162", 1/16" BRASS SHEET, C-353

Ratchet Fly Cutter for the Escape Wheel

The shape of the escape wheel cutter is similar to that of a ratchet fly cutter. It can be made on a grinder. The method described here was learned from W.R. Smith and adapted somewhat based on the writings of John Wilding.

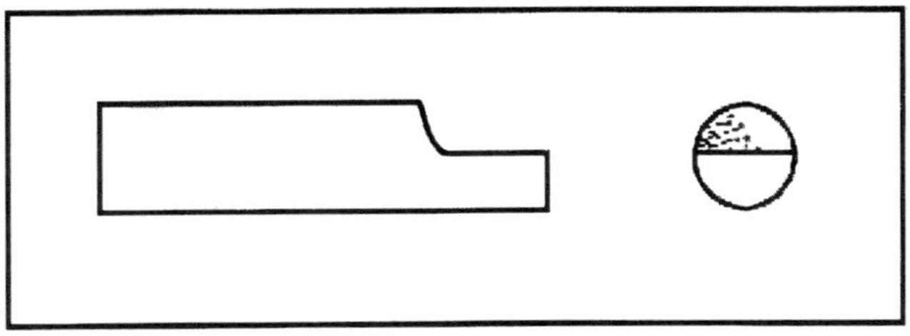

Fig. 39. First step in making the escape fly cutter.

The cutter is made from a piece of 1/4" drill rod which is between 1" and 1-1/2" long. As shown in Figure 39, begin by filing the end of the blank down to exactly the center line.

Now fasten the cutter blank in the square holder, which is described in Chapter 5. The key to setting up the grinder, as shown in Figure 40, is to place the square holder on a spacer. A piece of 3/8" thick

Fig. 40. The grinder set up for the cutter.

aluminum works well. This raises the cutter blank above the center of the grinding wheel, making two of the cutter clearances automatic. You don't have to think about them.

Figures 41 through 43 show the three steps in grinding the cutter. In Figure 41, the angle of 45° is placed onto the cutter. The angle can be checked with a protractor made for this purpose or by cutting out a 45° guide from heavy paper. In Figure 42 the side relief is added, then in Figure 43 the flat tooth bottom is ground.

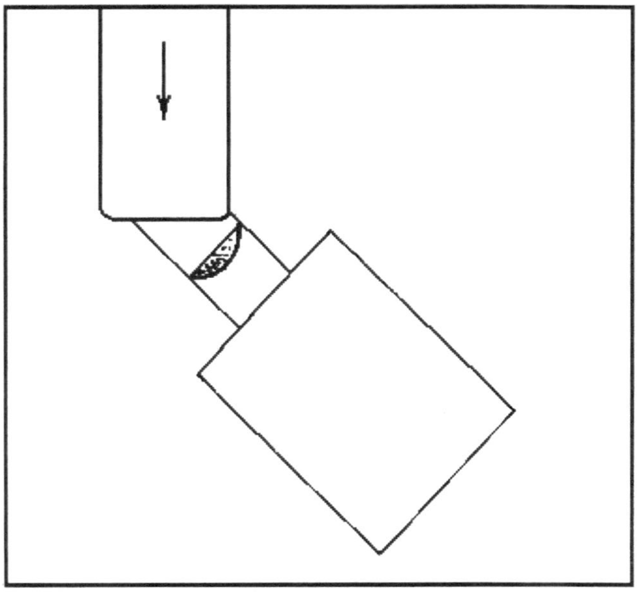

Fig. 41. Grinding the 45° angle onto the cutter.

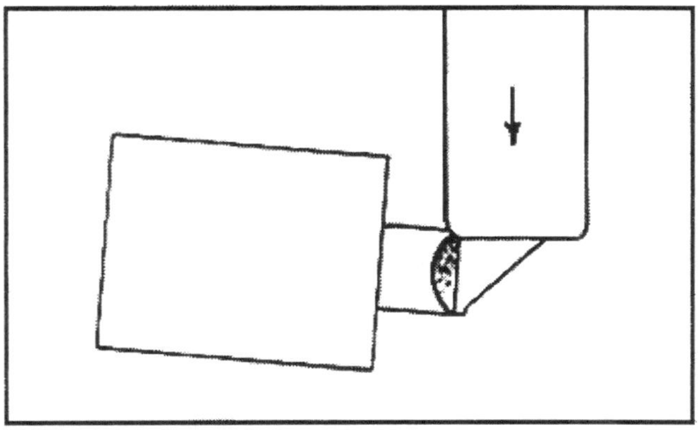

Fig. 42. Adding the side relief.

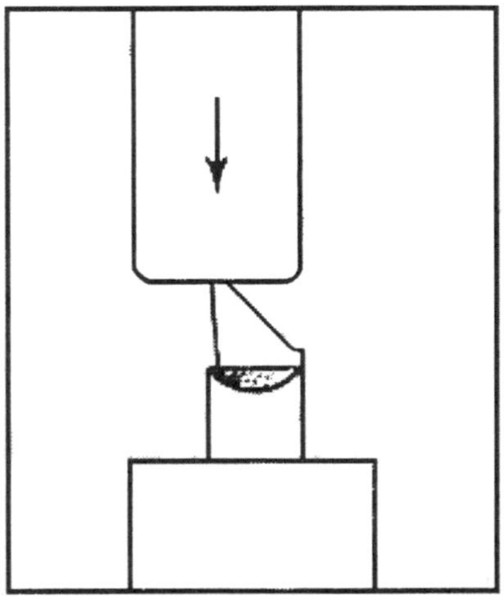

Fig. 43. Grinding the tip of the cutter.

Of these three grinding operations, the side relief (Figure 42) is the most difficult to visualize. In the view shown in Figure 44, the cutter would be rotating in order to take a cut from the wheel blank. Without the side relief, the entire radial edge of the cutter would rub against the side of the escape wheel tooth. The cutter would produce noise and vibration and might even bend the escape wheel teeth as they were being cut. The small amount of relief shown in the drawing allows only the tip of the cutter to touch the side of the escape wheel teeth. The drawing also shows that the side relief does not take away from the desired 45° cutter angle. The finished tooth would still look the same if the side relief were a little more or a little less than that shown in the drawing.

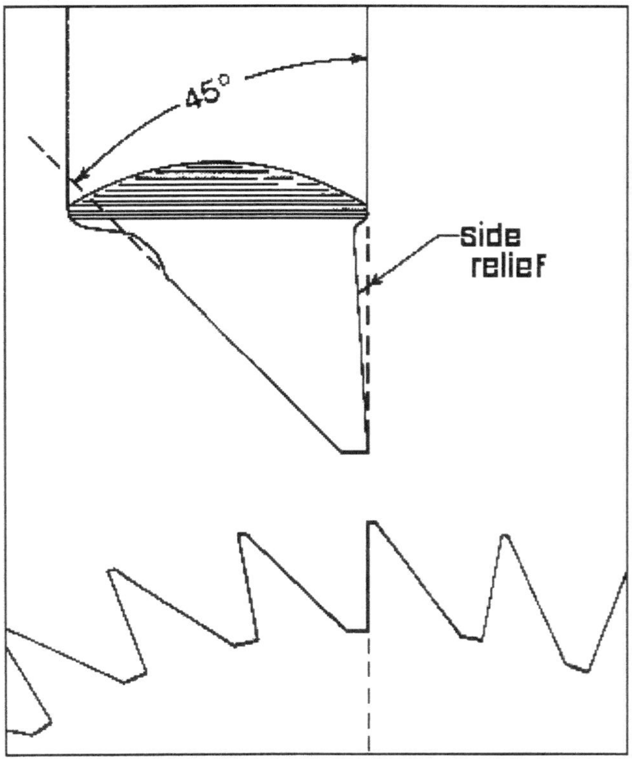

Fig. 44. Side relief on the cutter.

The completed cutter is shown still secured in the square holder in Figure 45.

The next step is to remove the cutter from the square holder and take off any burrs with a hand held India stone. Harden the cutter in the following manner. Heat the tip of the cutter slightly and dip it into powdered boric acid. This will prevent the formation of hard-to-remove black scale. To obtain enough heat to treat the 1/4" diameter cutter with an ordinary propane torch, place a fire brick in an upright position. Use tongs to hold the cutter close to the fire brick. Direct the flame onto the cutter until the end is cherry red. Quench immediately in water. The cutter can be reheated slowly, to temper it to a light straw color and make it less brittle, but I have left cutters of this type dead hard without suffering any broken cutters during the wheel cutting operation.

Polish the cutting edge opposite the relief edge with fine polishing paper. This step will greatly improve the smoothness of the escape wheel teeth. An unpolished cutter, even if the roughest burrs have been removed, will leave grooves in the wheel teeth.

The cutting procedure is similar to making train wheels with the Huckabee Wheel Making Fixture, covered in Chapter 3. However, some special details for the escape wheel will be helpful.

Make two brass escape wheel blanks and set them up on the Huckabee fixture. Set the escape wheel fly cutter on center as shown in Figure 46. The tip at the relieved edge is lined up with the center on the fixture. This setting will cause the backs of the escape wheel teeth to be cut radially, that is, pointing to the center of the escape wheel.

...text continues on page 23

Fig. 45. The cutter gripped in the square holder.

Fig. 46. Setting the escape tooth cutter on center.

FIG. 47. ESCAPE WHEEL

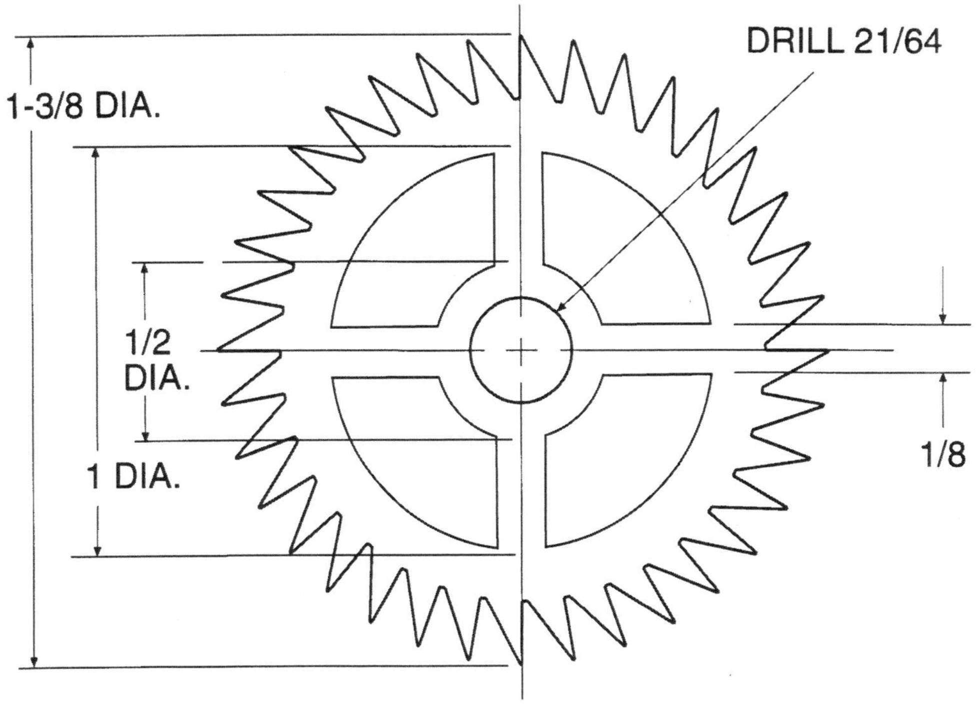

36 TEETH. ONE REQUIRED.
MATERIAL: BRASS SHEET, .032" THICK.
FOUR OR FIVE SPOKES MAY BE USED.

All the escape wheel dimensions are given in Figure 47, on the opposite page. After the edges of the escape wheel blanks are coated with layout dye, the actual cutting of the teeth can begin. The first tooth should be cut at a trial depth and made shallow. It is only when the next tooth is cut that the width of the tooth tip can be seen. The tip will be wide at first, then become narrower on successive cuts, as the cutter is adjusted deeper. The radial tooth face will not change in relation to the width of the tooth tip. The acting face of the escape wheel tooth will change, however, so keep that in mind as you make successively deeper cuts on the way toward the correct depth. The teeth are not cut to a sharp point, but must be allowed a flat tip which is not more than about .005" wide.

Once the correct depth of the cutter has been reached, cut all the teeth at full depth in one pass. Figure 48 shows the job being completed on a Sherline mill, but it can also be done on a Sherline lathe with vertical column or a Unimat-3 with the milling attachment. The Sherline setup allows the wheel blank to be moved through the path of the cutter, then reversed and brought back to the starting point under power. The index plate is then moved ahead one or more slots, depending on the count. As stated in the previous chapter, the Unimat-3 motor should be turned off after each pass, and the cutter turned to the side. The Huckabee fixture is then moved back to the starting position, where the index plate is set up for the next pass.

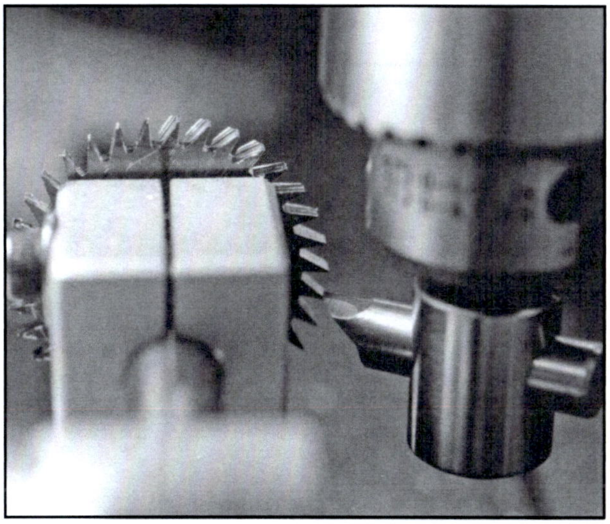

Fig. 48. Cutting the escape wheel teeth.

Fig. 49. The finished escape wheel in the movement.

Figure 49 shows the finished escape wheel when the movement is at a later stage of completion. Notice that the wheel has been crossed out, which is our next topic.

Crossing Out the Wheels

Several writers, including John Wilding and W.R. Smith, have described wheel crossing methods in detail. Some will consider it a tedious part of the construction process, but it is a necessary one. Wheel crossing is included here because this is the chapter on making the wheels for the clock, but (as I have already stated) many will want to leave the actual crossing out for later.

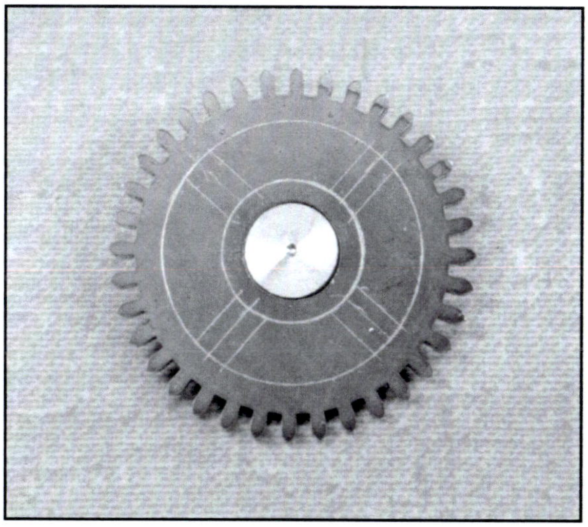

Fig. 50. Gear blank with spokes drawn in.

The wheels each have five spokes except for the hour, minute and escape wheels, which have four spokes each. Refer to Figures 35 through 38 and 47 for suggested spoke dimensions.

The first step in crossing out a wheel is to coat the surface with layout dye. In the next step, an idea from W.R. Smith works quite well: turn a button center from brass rod. As shown in Figure 50, the button fits the center hole. The button center can be removed and inserted with the fingers, but it will not fall out when the wheel is picked up. At the time the button center is made on the lathe, a small depression is made in it with a center drill. After parting off the button center, insert it in the wheel hole as shown in Figure 50. You now have a center for dividers. Scribe the two circles on the wheel, for the hub and the base of the toothed portion. Draw in the four or five spokes with a sharp scriber, making the marks clear but not so deep that it will be hard to polish them out later.

Drill a small hole through each of the four or five waste areas that will be cut away as the spokes are formed. Thread the jeweler's saw blade through a hole, then cut out the wedge shaped section. It is important to cut as close as possible to the lines, to minimize the filing you will have to do.

Figure 51 shows a triangular file being used to finish a spoke section. Another style, the barrette file, is essential for getting into corners. Speed improves with experience, but the filing task is still a long one, an hour or so for each wheel.

The wheels are not lacquered. Create a smooth line finish by stroking them across a piece of Scotch-Brite abrasive. This finish is in keeping with the style of movement and is more appropriate than a highly polished finish. As much as possible, avoid touching the brass with your fingers after the wheels are finished. If they become marked from handling, remember to touch up with the abrasive before final assembly of the movement.

Fig. 51. Crossing out a wheel.

5
MAIN WHEEL ASSEMBLY

Fig. 52. The main wheel assembly.

The main wheel assembly, shown in Figure 52, is the most complex subassembly in the movement. It is made up of the main wheel, winding arbor, ratchet wheel, click, rivet, click spring, click spring retainer, and tension washer. The mainspring, which is hooked onto the winding arbor, is the only purchased part of the assembly.

Figure 53 shows an original American main wheel assembly next to the winding arbor, main wheel, and ratchet wheel for our project. The American main wheel assembly has the ratchet wheel staked onto the winding arbor. The main wheel is free to turn on the arbor, although it is held under tension with a large brass washer and is controlled by the ratchet and click. The ratchet allows counter-clockwise winding and clockwise running in this clock. The click subassembly is riveted to the main wheel.

The repairer who has made an entire main wheel assembly can later make another complete assembly or any part of it for a repair. When you see the results of the desperate filing, hammering and soldering efforts that some repairers make to get a worn or broken click and ratchet to function again, you can appreciate that it is sometimes necessary to make a new part. Such work is not inexpensive, however, and not every customer is willing to pay the cost of custom made replacement parts.

In Chapter 4, we cut the teeth on the main wheel. I made the suggestion that wheels are better left to be crossed out toward the end of the project, just before they are cemented to the wheel collets. With the main wheel, there is no such choice. The wheel must be crossed out before the click is riveted in place and the wheel is staked onto the winding arbor, which is all covered in this chapter.

Ratchet Wheel

The ratchet wheel is shown in Figure 54. Our overall method in making the main wheel assembly is to make the ratchet wheel first, then the winding arbor, which should be given a straight knurl. If the fit is

Fig. 53. New parts (left); an old main wheel (right).

Fig. 54. The ratchet wheel.

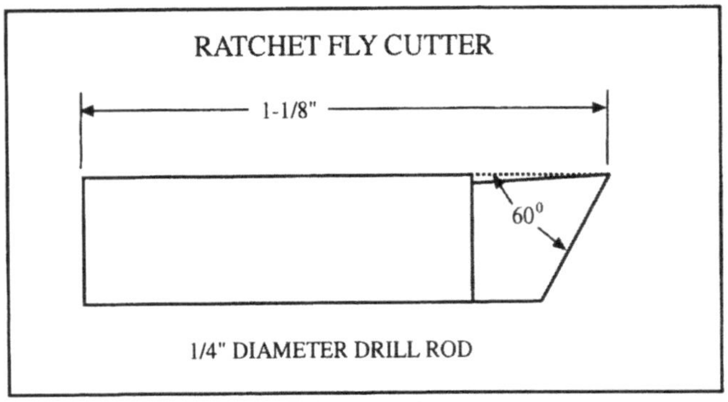

Fig. 55. Ratchet fly cutter.

too tight, it is easy to remove some material from the winding arbor to make the fit correct for the ratchet wheel to be staked onto it.

Going step by step, the first task is to produce the fly cutter for making the ratchet wheel. The procedure is one I learned from W.R. Smith, and it is the

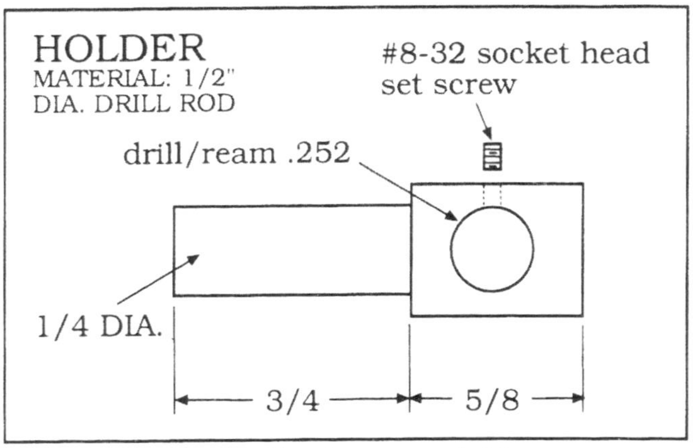

Fig. 56. Plan for a cutter holder.

same basic approach that we used in Chapter 4 to make the escape wheel fly cutter. The main difference is that the cutter angle is 60° instead of 45°. The ratchet fly cutter is shown in Figure 55.

The 1/4" cutter requires a holder. Figure 56 shows the holder made for this project. It was based on a design which was described by John Wilding in a magazine article. The holder can be made on a miniature lathe.

Proceed to make the ratchet wheel from brass rod. The drawing on the opposite page gives the dimensions. A step-by-step procedure follows. I used the Unimat-3 for making the ratchet wheel. The tools and accessories used are listed below:

division plate	*male tailstock center*
ratchet fly cutter	*brass rod*
milling attachment or vertical column	*cutter holder*
drill chuck	*parting tool*
3-jaw chuck	*lathe tool bit*

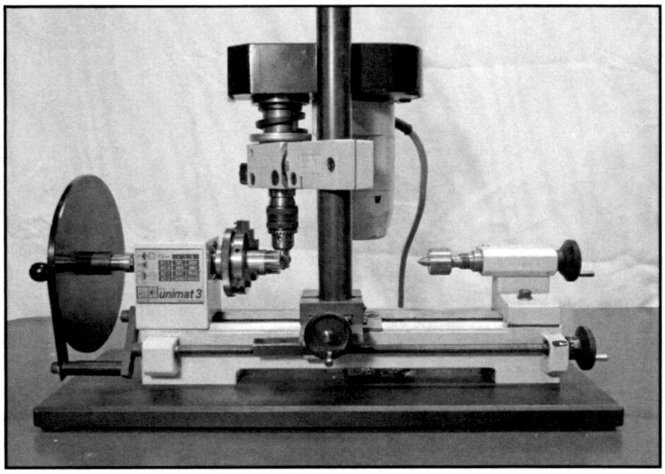

Fig. 57. Unimat-3 set up for fly cutting.

Figure 57 shows the Unimat milling column installed on the cross slide. The drive unit has been removed from the headstock and attached to the milling column. A male center is installed in the tailstock and is used during several of the steps in making the ratchet wheel. The division plate is the large disc at the left end of the headstock spindle. The brass rod is held in the 3-jaw chuck and is moved to the next tooth position after each cut.

Step 1: Preparing the Ratchet Blank

Dimensions for the ratchet wheel are shown on the opposite page. The brass rod selected for the wheel blank is 7/8" in diameter. Cut off 1-5/8" to 2" of rod to use as the ratchet wheel blank. It is a good

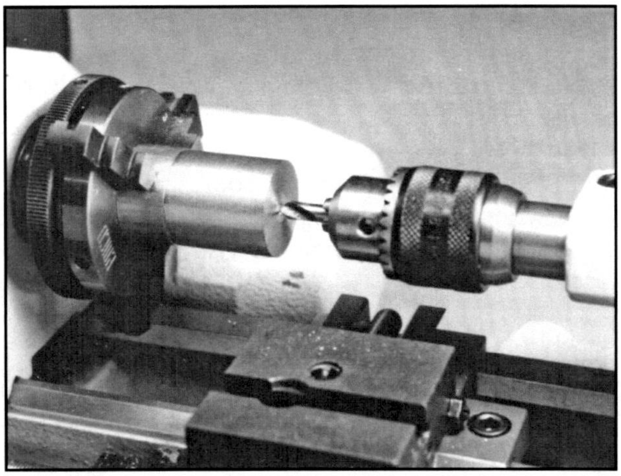

Fig. 58. Drilling the blank to receive a center.

idea to set up the Unimat as shown in Figure 57 for a "dry run", just to see how the pieces will fit for fly cutting. With the cutter in place, you can visualize its swing and judge how much clearance is needed between it and the 3-jaw chuck. It is also necessary

FIG. 59. RATCHET WHEEL

MATERIAL: 7/8" DIAMETER BRASS ROD
ONE REQUIRED

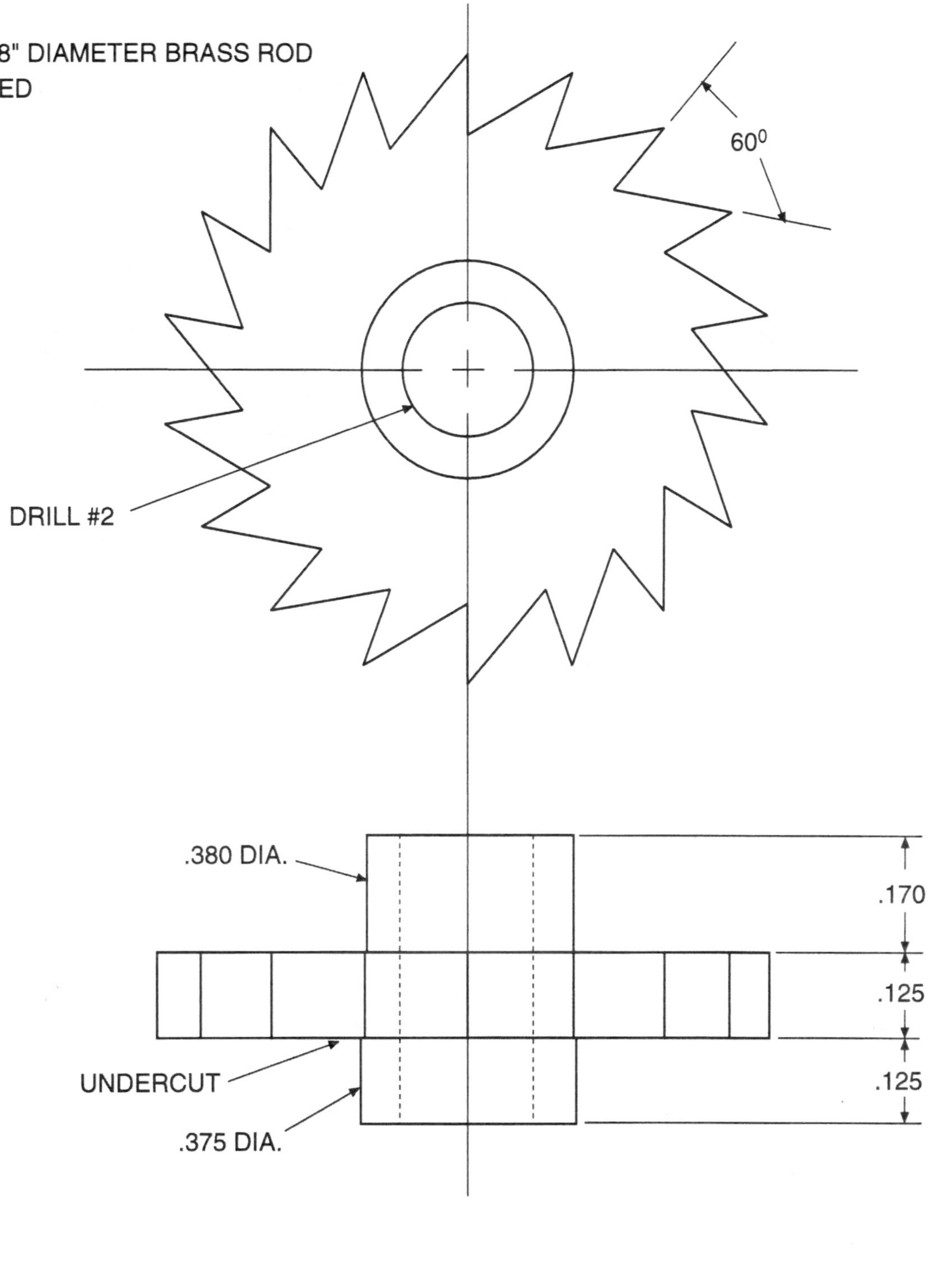

FLY CUT 18 RATCHET TEETH.
THE MAIN WHEEL FITS OVER THE .375" DIA. REAR HUB. UNDERCUT RATCHET SO MAIN WHEEL CONTACTS IT AT OUTER EDGE.

to allow a length of brass material for the front hub of the ratchet wheel.

Remove the milling column from the lathe and install the drive unit back onto the headstock. Mount the ratchet blank in the 3-jaw chuck and face off the end with a lathe tool bit. To retain concentricity, it is essential to avoid removing the ratchet blank from the 3-jaw chuck until the piece is parted off at completion.

The ratchet blank needs to be drilled to receive a male center for support. A center drill is used in Figure 58 to make the hole for the center. Install the

Fig. 60. Machining the front hub of the ratchet wheel.

center in the tailstock, bring the tailstock into position, and then crank the center in firmly, without forcing it. The point of the center does not touch the brass; only the conical portion makes contact. A dead center, which does not rotate, must be lubricated and checked from time to time for looseness in the ratchet blank. On the other hand, the live center rotates with the work and does not need lubrication.

Turn the ratchet blank down only enough to true up the piece in the 3-jaw chuck. Less than .010" should be removed from the 7/8" diameter. Proceed as shown in Figure 60 to machine the .380" diameter front hub. A small amount of extra length, per haps .025", can be left on the front hub, to be faced off later.

Step 2: Setting up the Cutter

Set up the Unimat-3 for milling, but do not remove the 3-jaw chuck or loosen the ratchet blank which it holds. Install the milling column on the cross slide and add the milling attachment. Transfer the motor drive unit from the headstock to the mill-

Fig. 61. Setting the cutter on center.

ing head. Loop the longer drive belt directly from the motor pulley to the milling spindle pulley, bypassing the countershaft pulley. This sets the highest spindle speed when the two-speed motor is switched to the higher setting. Screw the Unimat drill chuck onto the milling spindle nose and fasten the cutter holder in the drill chuck. Insert the ratchet fly cutter and adjust it by eye so that the flat side of the cutter is facing the ratchet blank and is set exactly in a vertical plane. Tighten the cutter in place and do not loosen it again during the job. Use the longitudinal and cross slide handwheels to bring the milling head to the point where the cutter is near the end of the ratchet blank. Loosen the clamping screw on the milling head and adjust the head so the cutter is near the center of the ratchet blank. Now insert the Unimat hand lever (which is also used for drilling) and carefully set the cutter height on center, as shown in Figure 61. The tailstock center has been moved away, and the center hole used as the reference for the cutter placement. Once the adjustment is made, tighten the clamping screw securely and remove the hand lever from the milling attachment.

Having set the cutter height, turn the handwheels to bring the cutter position closer to the actual cutting position. Tighten the gib on the cross slide to prevent the cross slide from moving accidentally during the fly cutting operation. The gib should be tighter than for normal turning work, but free enough to allow the handwheel to be turned without forcing it. To adjust the gib, refer to Figure 62. Loosen the two hex nuts on the side of the cross slide. Adjust the two set screws, checking the stiffness of the cross slide handwheel after each adjustment. When you are sat-

Fig. 62. Check the Unimat gib and saddle screw.

isfied with the action of the cross slide, tighten the hex nuts again to maintain the gib adjustment.

Also check the saddle screw, which is identified on the lathe in Figure 62. If the screw is loose, a large amount of play results in the saddle; if it is tight, the handwheel will not move at all.

Figure 62 shows a piece of tape on the lathe bed to mark the end of the fly cutter travel to the right, away from the headstock. After each cut, the lathe saddle is brought back to touch this tape, saving the operator from going too far or not far enough. This is especially helpful because it is difficult to judge the position of the whirling cutter. Another piece of tape is placed on the lathe bed on the other side of the cross slide, to mark the furthest travel toward the headstock. It is a safeguard against bringing the fly cutter into the 3-jaw chuck.

Fig. 63. The Unimat division plate.

Step 3: Dividing

Attach the division plate to the headstock end of the lathe. This is the Unimat accessory shown in Figure 63 with ten rings of holes ranging from 62 to 120 divisions. The ratchet wheel for the movement has 18 teeth, so the ring of 72 holes is used. There will be four holes counted each time the latch arm is seated to establish another tooth location. Double check the setup to make sure the latch arm is set in the ring of 72 holes.

Step 4: Fly Cutting the Ratchet Teeth

Although the cutter height has already been set (Figure 61), the depth of the cut must now be determined with trial cuts. When the cross slide handwheel is backed out, the cut is made deeper. This is because the cutting takes place on the back side of the ratchet blank. Coat the blank with layout dye. Set up the cutter to make a light pass first. The objective will be to cut two teeth initially, adjusting the depth with repeated cuts until a thin line, still marked with the layout dye, remains between them. In this way the full diameter of the ratchet wheel is maintained. The teeth will not have a knife-sharp edge.

Fig. 64. Fly cutting the ratchet teeth.

Make sure that all screws and adjustments are secure, and that you are wearing safety glasses. Remember that the cutter must be clear of the work at the start of each pass, with the lathe saddle at the right-side position shown in Figure 62. Turn the motor on high speed. Move the longitudinal handwheel slowly, especially after the cutter makes contact with the blank. Fly cutting is difficult to visualize until

Fig. 65. Turning the ratchet wheel rear hub.

you have tried it and realize that each rotation of the cutter cuts only a thin chip of brass from the blank. That is how teeth can be cut full depth in one pass. The chips fly everywhere, making eye protection especially important.

When the cutter has been advanced to the end of its intended travel, reverse the longitudinal handwheel to move the lathe saddle back to the starting point. With the Unimat-3, you may hear some additional cutter noise when the handwheel is reversed, because the saddle shifts slightly and allows the cutter to remove material on the way out. This action will not harm a ratchet tooth cutter, although it will break a gear tooth fly cutter.

With the motor still running, index through four holes on the division plate to reach the next cutting position. Make a trial cut at the same depth as the first tooth. Return the cutter to the starting point, shut off the motor, and look at the width of the dye coated space between the two teeth just cut. Adjust the cut deeper, and recut the same teeth until the depth is right, with just a thin line several thousandths of an inch wide showing. Once the depth is set, cut all the remaining teeth to full depth in one pass around the blank. The Unimat motor will heat up, so check frequently and give it a rest to cool off.

Figure 64 shows the appearance of the blank after most of the fly cutting has been done. The tailstock center is not in place because it was in the way of the cutter and did not seem needed during the fly cutting of this ratchet.

When the teeth are finished, use the parting tool to turn the rear hub of the ratchet wheel. A dead center is shown in use during this procedure in Figure 65. This hub must be the exact diameter (.375") over which the main wheel will fit.

Remove the tailstock center and drill out the center hole, as shown in Figure 66. The drill is a #2 size. Finish the ratchet wheel by parting it off at the correct length, at the rear hub. Refer to Figure 59, which shows that a slight undercut on the ratchet wheel will make it fit the main wheel better.

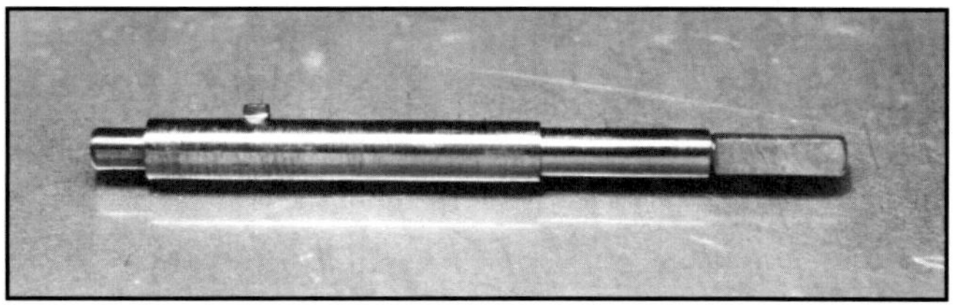

Fig. 67. The winding arbor.

Winding Arbor

Figure 67 shows the winding arbor, which is made from a piece of 1/4" drill rod. It can be made on any miniature lathe equipped for turning between centers. This method is useful whenever a piece is made up of several diameters which must remain concen-

Fig. 66. Drilling the hole through the ratchet.

Fig. 68. Machining between centers.

tric. We will also show a way to accurately file the winding square.

Begin by cutting off a piece of the drill rod about 3-1/8" long. Mount the drill rod in the three jaw chuck, with only a small length exposed. Face off the end and center drill it to receive a center. Make sure the drilling is deep enough to prevent the point of the center from making contact with the drill rod; the conical portion of the center makes the contact instead. Remove the piece from the chuck, turn it around, and prepare the other end in the same way. The piece can now be mounted between centers as shown in Figure 68. The photo shows how the faceplate is screwed onto the spindle in place of a chuck. The lathe dog is then tightened onto the work piece. The faceplate rotates the lathe dog, which in turn rotates the work piece. The tailstock center must be lubricated unless it is a live center. Turning between centers allows you to remove the workpiece and turn it around end-for-end any number of times, without losing the concentricity. On the negative side, it sometimes takes extra time to set up the lathe to make the required cutting pass, since the tailstock limits the travel of the cross slide. Figure 69 shows the appearance of the winding arbor after the front pivot has been machined between centers.

Fig. 69. The front pivot machined between centers.

Figure 70 shows the dimensions of the winding arbor, including the mainspring hook.

...text continued on the next page

FIG. 70. WINDING ARBOR

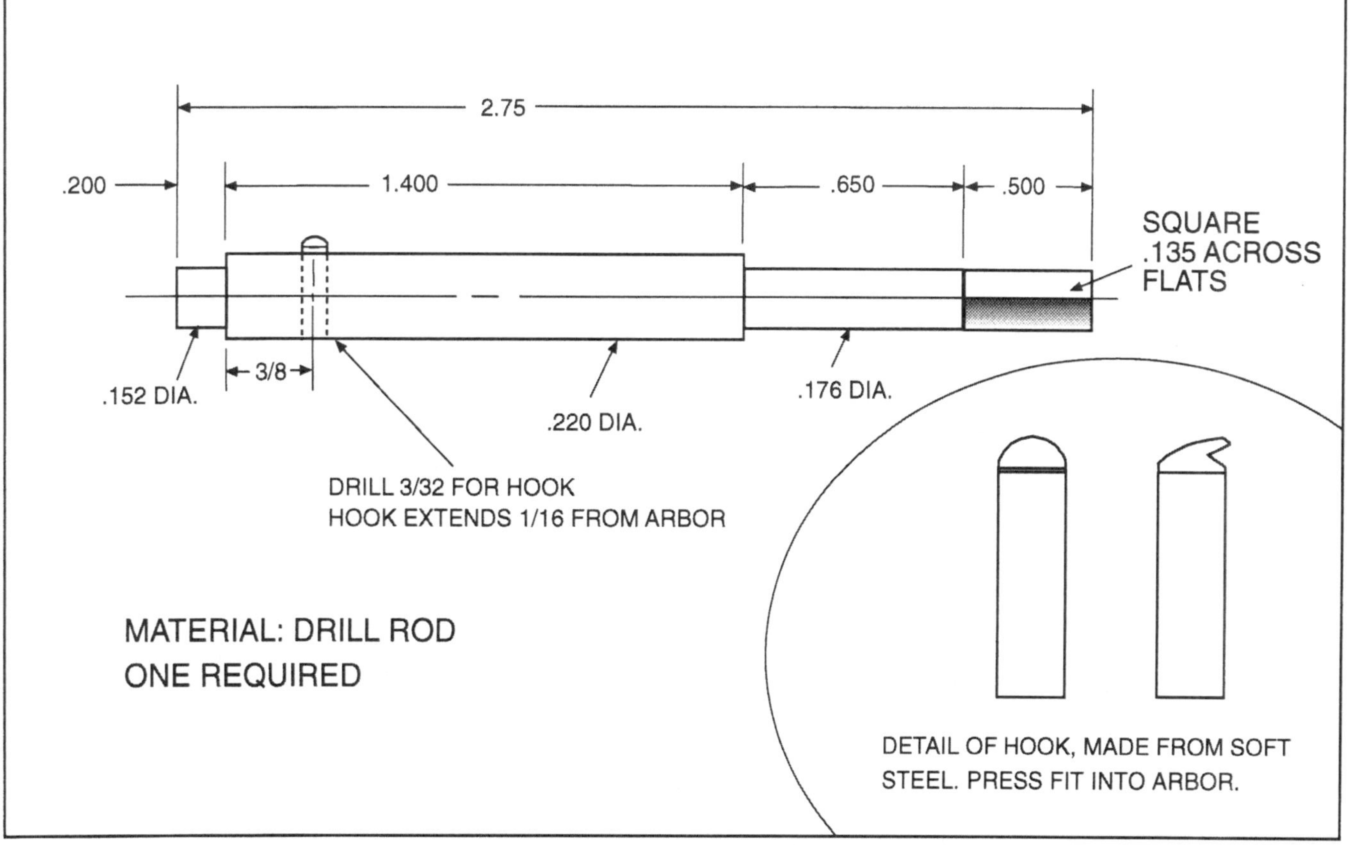

Filing the Winding Square

Some lathes are fitted with a filing rest which helps to produce an accurate result. However, it is possible to accomplish the filing in another way, by making a square fixture. Take a piece of 3/4" square steel stock about 1-1/4" long and drill it end to end with a 1/4" diameter hole for the winding arbor. Cross drill

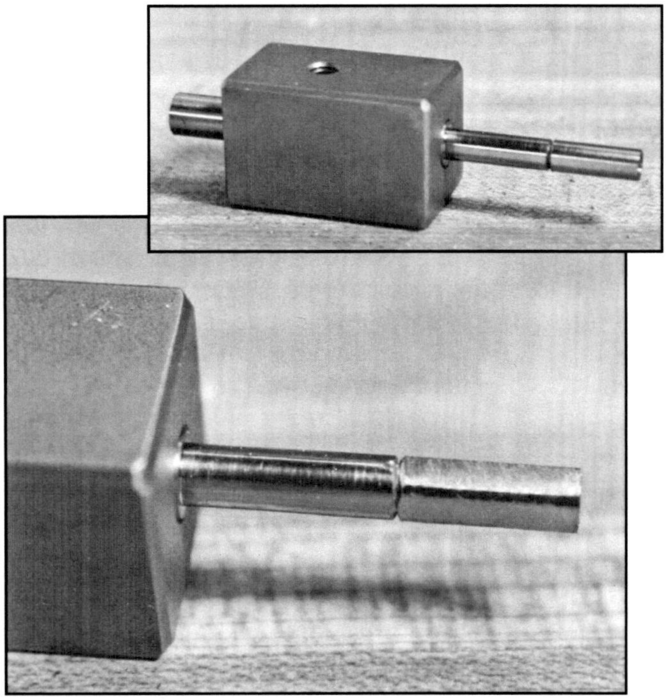

Fig. 71 and 72. Two views of the square holder.

the square piece for a set screw to clamp down on the winding arbor to hold it in place. The results are shown in Figures 71 and 72.

Once the winding arbor has been placed in the square fixture and the set screw has been tightened, place the fixture in the vise. Coat the arbor with layout dye. Keeping the file horizontal, file a flat on the top, 1/2" in length. (I believe it was John Wilding who suggested turning a groove in the arbor to mark the end of the square portion, which makes a neater filing job.) After some initial filing, loosen the vise and turn the fixture to another side. File again, watching the width of the dye-marked edges remaining at the corners of the square. File until you leave just a thin line between each of the flats. Check across the flats with a micrometer or calipers, which will show the widths across the finished flats to be equal.

Turning the Rear Pivot

The arbor can be removed from the square fixture and be placed between centers again in the lathe. Figure 73 shows the rear pivot being turned to .152" diameter. A parting tool can be moved side to side as

Fig. 73. Turning the winding arbor rear pivot.

Fig. 74. Drilling the hole for the hook.

Fig. 75. Shaping the hook.

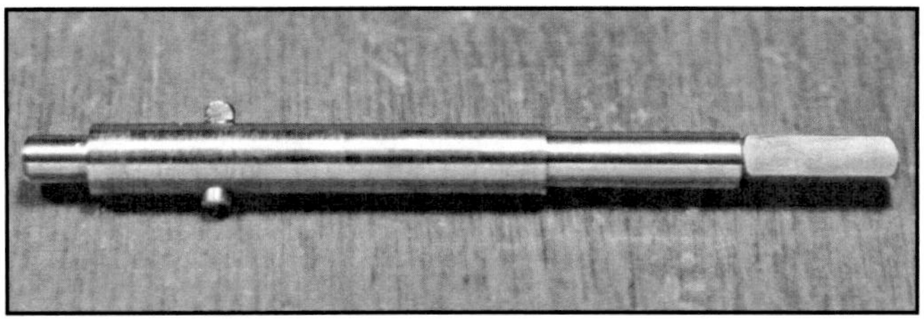

Fig. 76. Another view of the winding arbor.

Fig. 77. Drilling the main wheel.

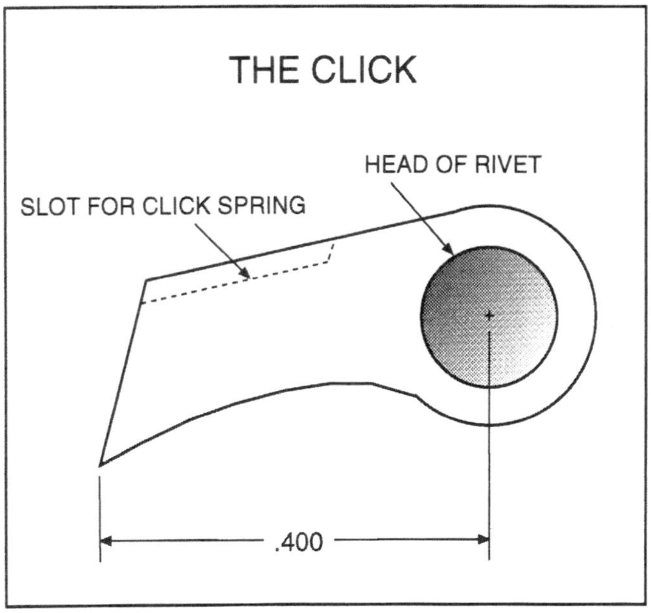

Fig. 78. The most important dimension of the click.

shown, taking light cuts. The center portion of the winding arbor can be machined in the same way, reduced from 1/4" to .220" in diameter.

The Mainspring Hook

The hole is now drilled 3/32" for the mainspring hook, as shown in Figure 74. The hole is shown being drilled in the Unimat-3 with the arbor held in the machine vise. (Dimensions and location of the hook are shown in Figure 70.) Figure 75 shows the hook being shaped. The hook is turned or filed down to fit in the hole in the arbor. The piece should be a tight fit to hold the mainspring securely.

Figure 76 shows the winding arbor almost finished. The working end of the hook should extend about 1/16" from the arbor. The other end should be filed flush with the arbor. The ends of the winding arbor will show the holes used for center turning, unless this material is faced off the arbor.

Click and Rivet

Figure 77 shows the main wheel being drilled for the click rivet. You will find the locations and sizes of all four holes which are needed for the rivet and click spring shown in Figure 80. Note that the location of the hole for the rivet is for reference; check the fit of your click before drilling.

As you begin to make the click, you will see that the exact shape could vary somewhat. One approach to the making of this part is to cut a proposed shape of click from paper and place it on the main wheel, along with the ratchet wheel, for a trial fit. Cut out this shape and trace it onto a piece of 1/16" thick clockmaking brass which has been coated with layout dye. Drill a #34 hole in the brass first, before it is cut out. Figure 79 shows the right way to shape the end of the click, as well as two ways it is sometimes

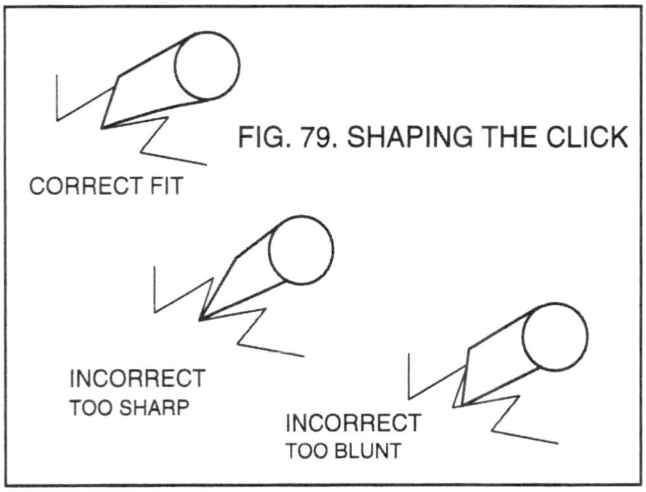

Fig. 79. Correct and incorrect shapes for the click.

FIG. 80. HOLE LOCATIONS - MAIN WHEEL

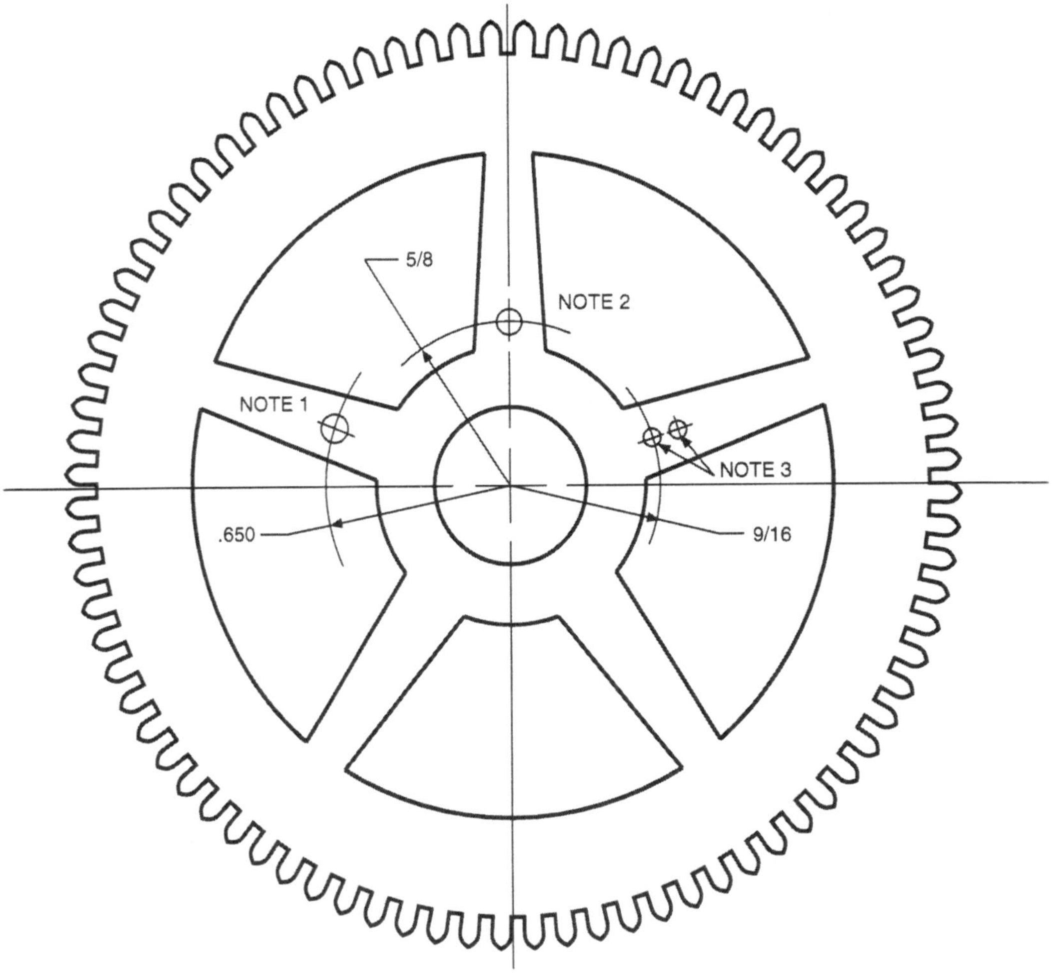

NOTE 1. HOLE FOR CLICK RIVET. DRILL #44. FIT PARTS AND ADJUST LOCATION BEFORE DRILLING.

NOTE 2. HOLE FOR END OF CLICKSPRING. DRILL #55.

NOTE 3. HOLES FOR U-SHAPED WIRE TO HOLD CLICK SPRING. DRILL #64.

SEE CHAPTER 4 FOR OTHER DETAILS OF MAIN WHEEL.

FIG. 81. CLICK RIVET

MATERIAL: DRILL ROD

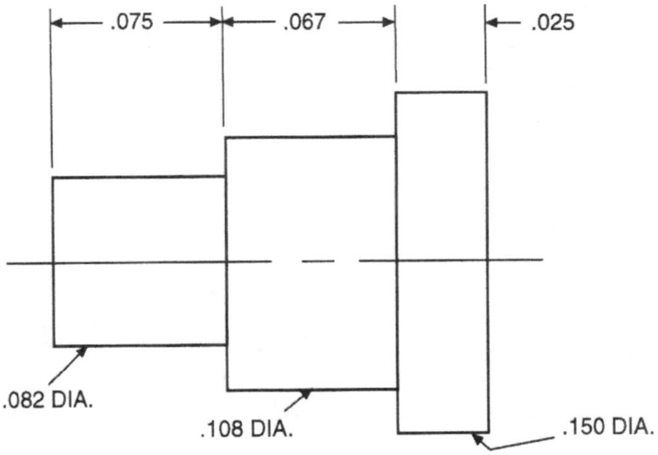

Fig. 82. Fit the click to the ratchet wheel before drilling.

done incorrectly. A good fitting click is essential for a winding mechanism that is safe and dependable. The click spring cannot compensate for a click that fits poorly.

Proceed to cut the click shape out of the brass with a jeweler's saw. For cutting, place the brass sheet on a piece of board with a "V" notch cut in it. Move the saw in a vertical plane, with the blade inserted so it will cut on the down stroke. Cut as close as possible to the drawn line, but leave some extra material at the tip of the click so that it can be fitted to the clickwheel.

File the click to final shape. Place it over the rivet hole in the main wheel to determine how the tip should be filed for the best fit. Give the finished click a smooth line finish. Use small files or a jeweler's saw to make the slot in the back of the click to receive the click spring. Look ahead to Figure 91 for the appearance of the finished click.

Fig 83. The main wheel crossed out and drilled for the rivet, click spring and U-shaped retaining wire.

The rivet, shown in Figure 81 (opposite page, bottom), is made from drill rod. The click should fit freely on the rivet, with no tendency to bind. Fit the ratchet wheel partly onto the winding arbor, and fit the rivet and click to the wheel, all very loosely so they can be taken apart again. With the click oriented up, it should operate even without a click spring, when the arbor is turned by hand. It is possible to see whether the click fits the ratchet wheel correctly. The point of the click can be filed to fit the ratchet tooth shape, as long as there is not too much material removed from it. It should be remembered that it is not possible to change the shape of the click after it is riveted to the main wheel. Figure 82 shows the wheel at this stage, with the click being tried on the ratchet wheel.

Drill the other three holes in the main wheel for the click spring. Figure 80 gives the dimensions and locations of the holes. Next, cross out the main wheel to look like Figure 83. (Suggested spoke dimensions for the main wheel are shown on page 16.) The crossing out should be done now, as it will not be possible after the parts are riveted to the wheel.

Fig. 84. The click spring.

Click Spring

The click spring, shown in Figure 84, is made from 18 ga. (.048" diameter) brass spring wire, which is available from clock suppliers. Cut a piece about 4 inches long. One end needs a 90° bend to fit in the #55 hole in the main wheel. Begin to carefully shape the click spring to look like Figure 84. Flatten the other end into a blade shape by hammering it. This end will ride in the slot which is filed in the back of the click. Continue to bend the click spring without putting any sharp bends in it. The spring must form a curve which goes between two #64 size holes drilled

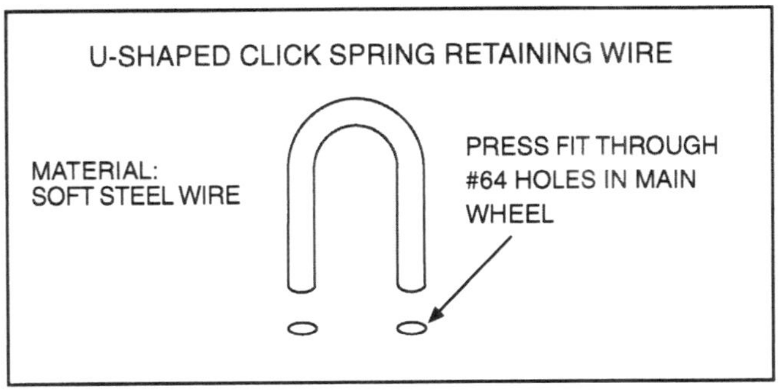

Fig. 85. U-shaped wire is made to fit over the click spring.

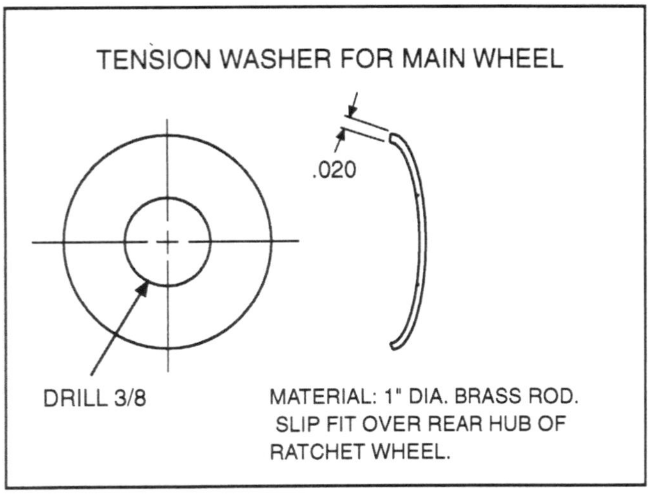

Fig. 86. The tension washer holds the main wheel in place.

Fig. 87. Method of placing a dome shape on the washer.

in the wheel. A small U-shaped wire is formed to go through these holes and capture the spring. When the spring has been formed to the desired shape, it will have enough tension to seat the click safely even when oriented upside down. Install the click spring in the #55 hole, lay it flat on the wheel, then press the U-shaped wire over it and into the holes. (See Figure 85.) Use Loctite to cement the U-shaped wire in place. Let it cure for a day (during which time the rest of the wheel assembly can be completed), then file off any excess wire from the underside of the wheel. This permanently attaches the click spring to the wheel, although it is still possible to adjust the tension of the spring later.

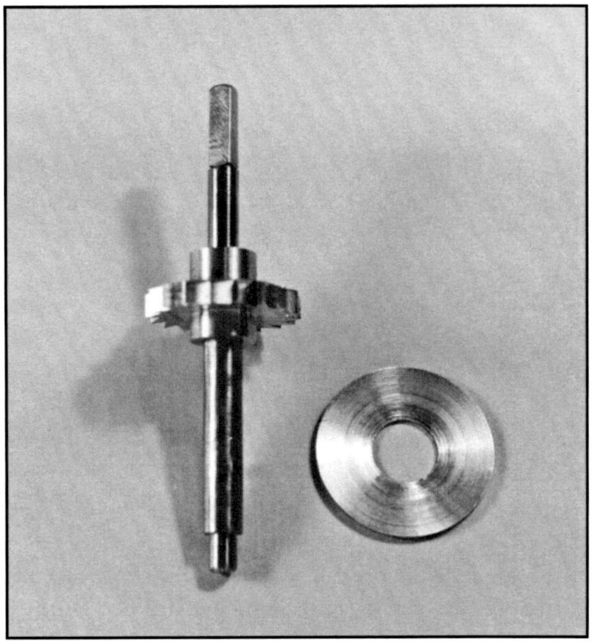

Fig. 88. Ratchet wheel, arbor and tension washer.

Tension Washer

The tension washer is dimensioned in Figure 86. After it is made from brass rod, the washer must be molded into a dome shape. Figure 87 shows a way of doing it. A 1-3/4" diameter steel ball is used in a vise to press on the washer, which is backed up by a piece of thick leather. The domed washer will place pressure on the wheel to hold it straight, yet allow the winding arbor and ratchet wheel to rotate.

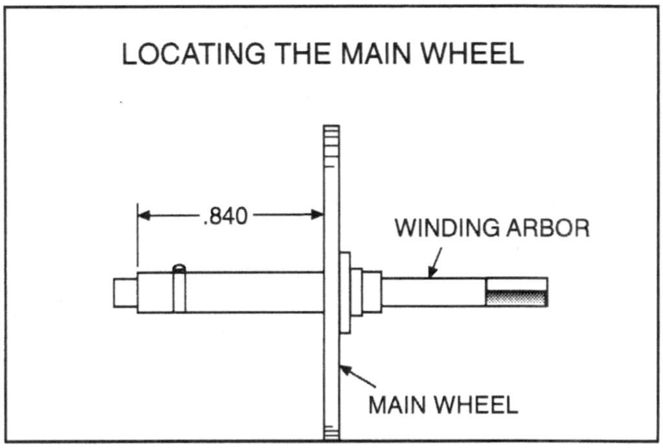

Fig. 89. Critical dimension for locating the main wheel.

Fig. 90. Rear view of the completed main wheel assembly.

Assembling the Main Wheel Parts

Figure 88 shows the ratchet wheel and winding arbor after the two parts have been staked together. The fit must be tight, so the winding arbor should be given a straight knurl in the area where the ratchet will be staked onto it. If a knurling tool is not available, several ridges can be gouged along the winding arbor if you take care not to bend it. If you take apart a scrap American main wheel, you will see that several ridges are enough to obtain the fit. Loctite is not recommended as an alternate to the knurling.

Figure 89 shows how far the ratchet wheel should be staked onto the arbor: when the main wheel is placed on the hub of the ratchet wheel, there should be .840" between the back surface of the wheel and the rear shoulder of the winding arbor. This dimension allows clearance for the mainspring to be added later. Next, fit the tension washer over the main wheel, and rivet the tension washer in place by striking the back end of the ratchet hub with a hammer and flat punch. It should not be necessary to use excessive force. The job should be easier than riveting some of the tension washers on old clocks.

Check the fit of the main wheel. The wheel should not be wobbly on the ratchet hub; on the other hand, it should not be too stiffly mounted to allow you to hold the wheel with one hand and wind the arbor with a key held in the other hand.

Figure 90 shows a back view of the completed main wheel assembly. Notice that the extra material has not yet been filed off the U-shaped retaining wire for the click spring. There is no reason to hurry with the filing, since Loctite adhesive sets up completely after about a day.

A close-up of the front of the assembly is shown in Figure 91, indicating the shape of the click spring, the way it fits in the click, and the fit of the click in the ratchet teeth.

The mainspring has been added in the Figure 91 photo. Install your mainspring now or wait until later in the construction (Chapter 10). As received from the supplier, the mainspring is retained by a thin wire with the ends twisted together. Before removing the wire and letting down the mainspring, first put on safety glasses. Grasp the spring firmly in the palm of your hand, while wearing a heavy canvas glove on that hand. Untwist the wire with pliers and allow the mainspring to slowly expand and unwind in your gloved hand. Do not release it suddenly.

Fig. 91. A close-up of the main wheel assembly.

Clean the uncoiled mainspring with clock cleaning solution or a safe solvent, then rinse and dry it thoroughly. Wipe on a mainspring lubricant with a clean cloth. Use pliers to form the inner end of the spring to a diameter which just fits over the winding arbor. Avoid kinking the spring. Slip the spring onto the arbor. When the hole in the mainspring end slips over the hook, you may hear it snap into place. The hook must be completely seated, rather than just barely catching the spring. Use a mainspring winder to rewind the mainspring to a diameter about the same size as it was received from the supplier. Put on a new retaining wire and twist the ends together. Let the spring down into this loop. As an alternative to hand winding your own wire, material sources #4 and #5 (page 2) sell a mainspring clamp which is made from round or flat steel. It slips onto a loop-end mainspring of the type used in this project. The diameter of the right clamp is about 1-7/8 inches.

It is a good idea to clean your new mainspring, as described above, because it was probably factory coated with a rust preventive, rather than a spring lubricant. However, if you do not have access to a mainspring winder, you may have to lubricate the new spring without first cleaning it.

6
PINIONS & ARBORS

The pinions in the movement, except for three to be discussed later in this chapter, are the lantern type. Most of the lantern pinions found in old clocks are made up of two disks, called shrouds, which are drilled to accept pins, called trundles. The shrouds are staked onto the arbor some distance apart, and the trundles are inserted between them to form a pinion like the one on the left in Figure 92. For new construction, the lantern pinion should be made from a one-piece brass body which is easier to make and assemble than separate shrouds. Figure 92 shows this type of pinion body on the right. The trundles are made from blue steel wire or music wire, depending on the method used to cut them.

Fig. 92. Lantern pinions.

Fig. 93. A pinion body is held in the 3-jaw chuck.

Making the Pinion Bodies

I decided to make the pinion bodies on the Unimat-3 rather than on Sherline equipment because of the particularly useful Unimat Indexing and Dividing Attachment, which will be described. Other equipment (such as that illustrated by W.R. Smith and others) does the job equally well if the setup includes an indexing and drilling attachment which can be offset from center. I will describe a Unimat procedure and leave it to clockmakers to devise the best way to drill the holes for the trundles using their own equipment. Dimensions for the lantern pinions are found in Figures 98 through 100.

Begin by cutting off a 1-1/2" long piece of brass rod for the pinion body. Install it in the 3-jaw chuck with about 3/4" extending. Face off the end, then coat the faced-off end with layout dye. Crank a lathe tool so the point is on center; it is possible to do this accurately, since the facing-off operation will have left the center clearly marked. Crank the tool outward by the measurement of the pitch radius, which is half the "bolt circle" of each pinion (Figures 98 through 100). With the point of the lathe tool touching lightly, turn the chuck by hand to scribe the pitch circle onto the dye-marked end of the rod.

Continue by installing a center drill in the tailstock drill chuck and center drilling the piece. Then drill a #43 center hole for a length that will go all the way through the pinion body. Now prepare a simple reamer of the type described by W.R. Smith. Take about a 2" length of the 3/32" diameter music wire which is to be used for the arbors. Grind four cutting edges starting about 1/8" from one end and tapering almost to a point. Remove any burrs from the cutting edges. Install this reamer in the drill chuck, oil it, and ream the center hole. You will be left with a hole which will be a perfect slip fit for the arbor material.

As shown in Figure 93, machine all the outer diameters for the pinion body. These dimensions are shown on the drawings, and some of them include

Fig. 94. Unimat Indexing & Dividing Attachment.

a collet for a wheel. A parting tool works best for this machining work. Remove the chuck from the headstock, but do not loosen the pinion body in the chuck, or the concentricity of the pinion will be lost.

Figure 94 shows the Unimat Indexing and Dividing Attachment, Item #150.320. The attachment is fastened to the cross slide with two T-nuts, and the 3-jaw chuck is screwed onto its spindle.

The attachment includes a 24-hole index plate which can be used to index the drilling locations for lantern pinions with 8 pins. A new plate must be made for the seven-hole escape pinion, since the manufacturer does not offer a plate for this accessory with a hole count divisible by seven. Figure 95 shows the dimensions of such a plate.

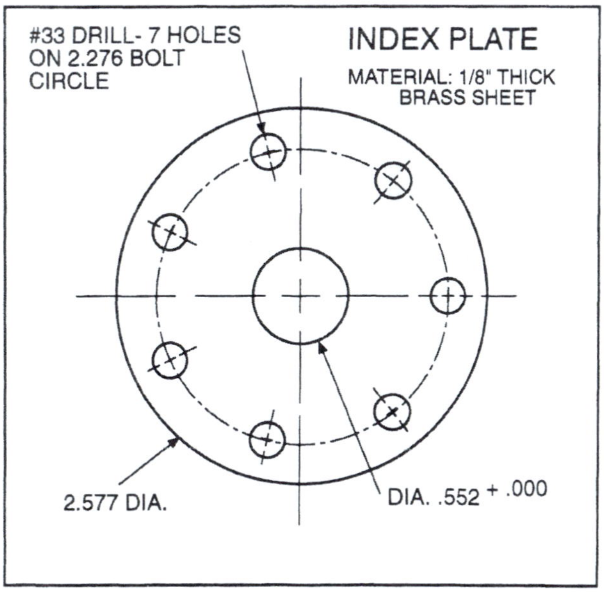

Fig. 95. Drawing of a 7-hole index plate.

Fig. 96. Center drilling the pinion.

Figure 96 shows the attachment, with the 3-jaw chuck mounted to it, on the Unimat cross slide. Install a small center drill and crank the cross slide until the point of the center drill touches anywhere on the scribed pitch circle. Turn on the machine and just touch the drill to start a center. Move the index-

Fig. 97. Drilling the pinion holes.

FIG. 98. SECOND AND FOURTH PINIONS

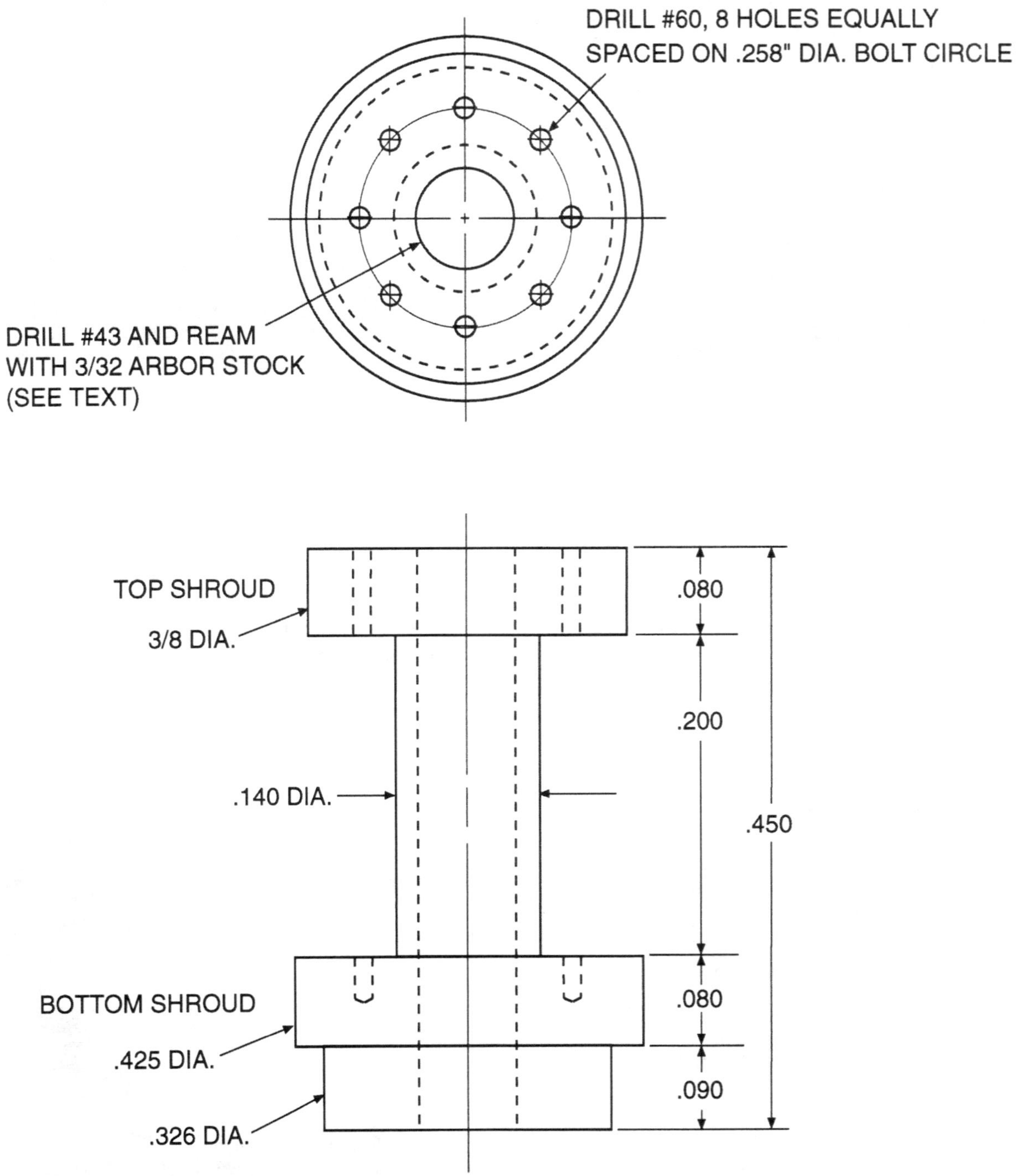

INDEX AND DRILL THE EIGHT HOLES THROUGH THE TOP SHROUD AND .050 INTO BOTTOM SHROUD.

PINS .040 MUSIC WIRE. FASTEN WITH LOCTITE.

FIG. 99. THIRD PINION

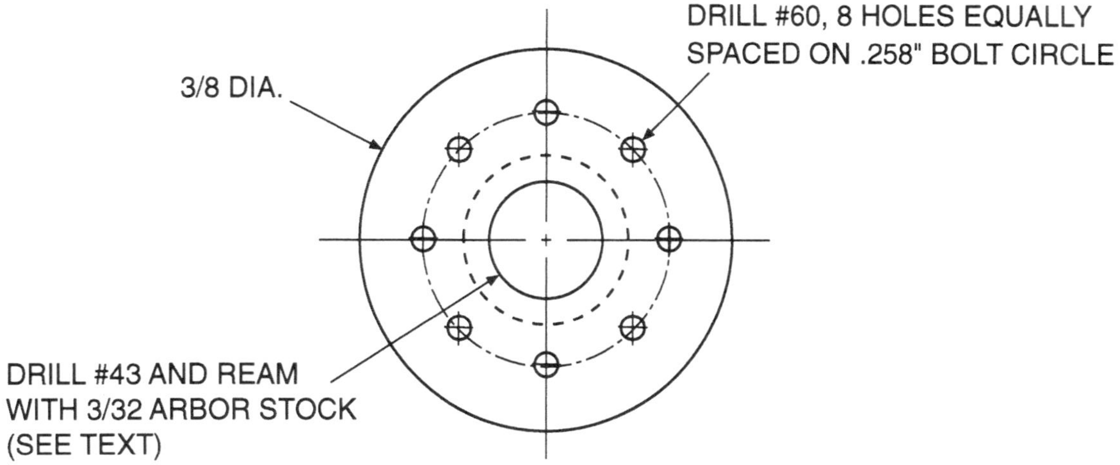

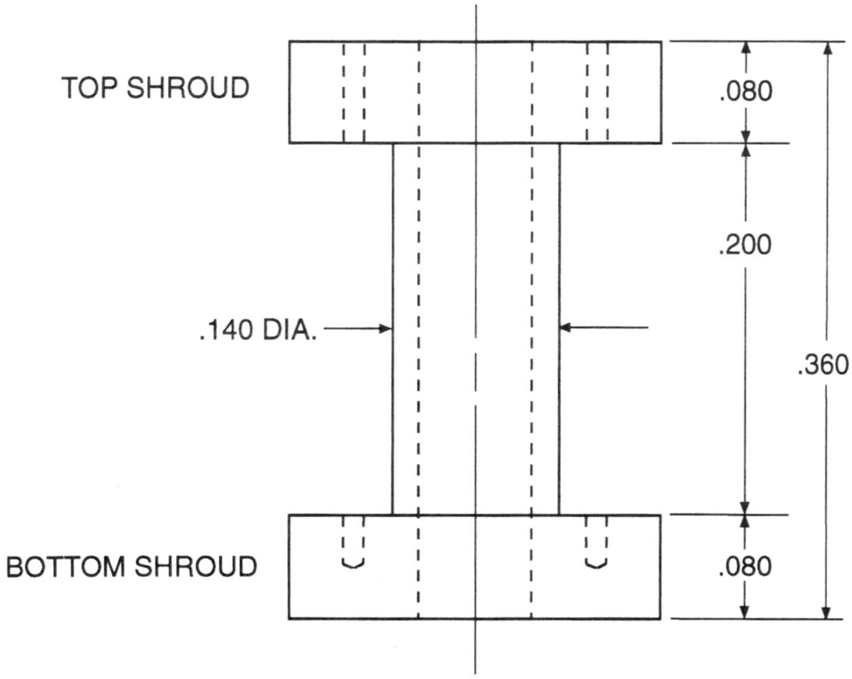

MATERIAL 3/8 DIA. BRASS ROD

INDEX AND DRILL THE EIGHT HOLES THROUGH THE TOP SHROUD AND .050 INTO BOTTOM SHROUD.

PINS .040 MUSIC WIRE. FASTEN WITH LOCTITE.

FIG. 100. ESCAPE PINION

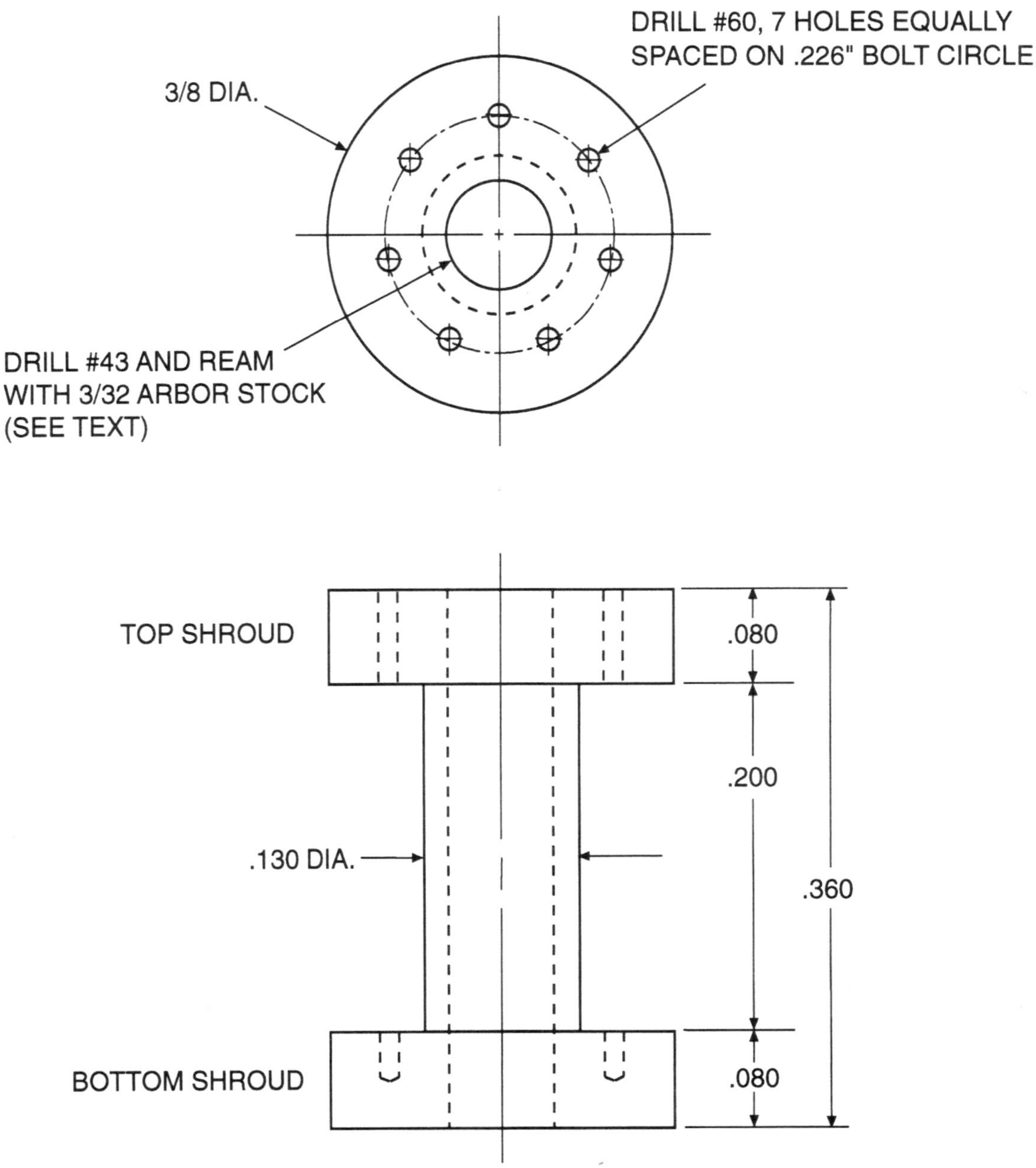

MATERIAL: 3/8 DIA. BRASS ROD

INDEX AND DRILL THE SEVEN HOLES THROUGH THE TOP SHROUD AND .050 INTO BOTTOM SHROUD.
PINS .040 MUSIC WIRE. FASTEN WITH LOCTITE.

Fig. 101. Pinion body.

ing attachment to the next position and make another center, and so on.

Now install a #60 drill in the chuck and drill the holes for the pins, as shown in Figure 97. As indicated in the drawings, these holes are drilled through the top shroud and about .050" into the bottom shroud. The Unimat drilling spindle is moved slowly so that the drill will not wander as the tip enters the lower shroud. This indexing and drilling procedure works well because there is no doubt that the circle around which the pin holes are drilled is exactly concentric with the center hole of the pinion. The only likely error is to scribe the pitch circle .001" or .002" too large or small.

After the holes are drilled, set the lathe back in the horizontal position and place the chuck onto the headstock spindle. Part off the finished pinion body. Figure 101 shows the appearance of the second and fourth pinions, which will each mount a wheel.

Installing the Pins

There are two ways in which pins, or trundles, can be made. One involves the use of blue steel wire for the pins. Blue steel comes in assortments of various diameters, and it can be difficult to obtain enough of a particular size. For our project, .040" diameter wire is needed. Fit a long piece of wire all the way through the pinion body until it seats in the bottom of the hole in the shroud. Mark the wire at a point which would make it extend slightly from the top shroud. Remove the wire and make a deep file cut at the mark. Place the wire in a small vise and snap it off at the marked place. Make all the pins in this way. Clean them with alcohol, then cement them in place with Loctite 609. Grind the pins off flush.

Another way of making the pins is to use music wire instead of blue steel. The .040" music wire is easy to obtain in long pieces, without having to buy an assortment of unneeded sizes. Music wire is very hard, but the advantage is that a wire shear can quickly cut off any number of required pieces, all the same length. An excellent wire shear I used was the Draucker Lantern Pinion Wire Shear (Figure 102), which is unfortunately no longer being made. It is

Fig. 102. Draucker Lantern Pinion Wire Shear.

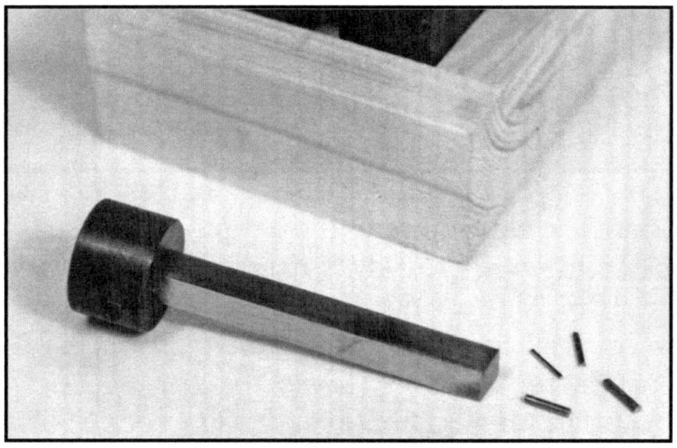

Fig. 103. Close-up of the cutter used in the shear.

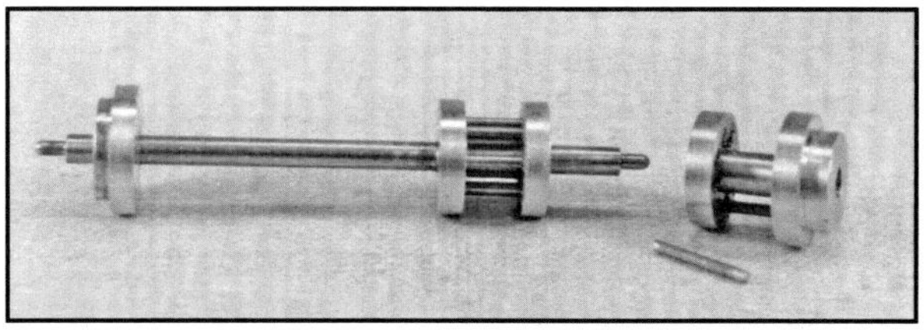

Fig. 104. Two pinions.

a two-piece steel block which uses a standard 1/4" lathe tool (Figure 103) ground at a 5° angle for shearing. A gauge screw permits the length of the wire to be set. Merritt's Antiques (supplier reference on page 2) offers a lantern pinion cutter which I have found effective.

The finished pinions can be inserted onto arbors. Although we haven't yet shown the making of the arbors, Figure 104 shows a preview.

Fig. 105. This fly cutter is used for:

- *Minute pinion, 10 teeth, .82 module*
- *Cannon pinion, 12 teeth, .82 module.*

Material: 1/4" diameter drill rod.
Harden and sharpen the cutter.
The dimensions are expressed in inches.

The round graver, Fig. 32, page 15, can be used to turn the radius onto the cutter.

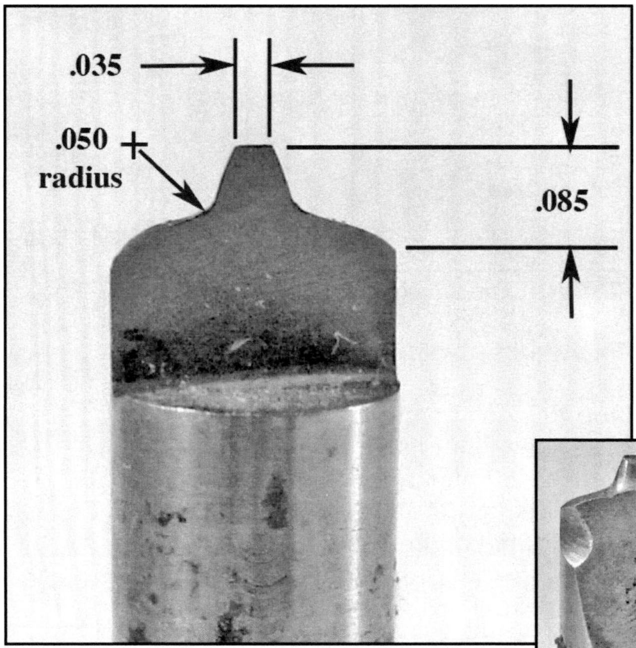

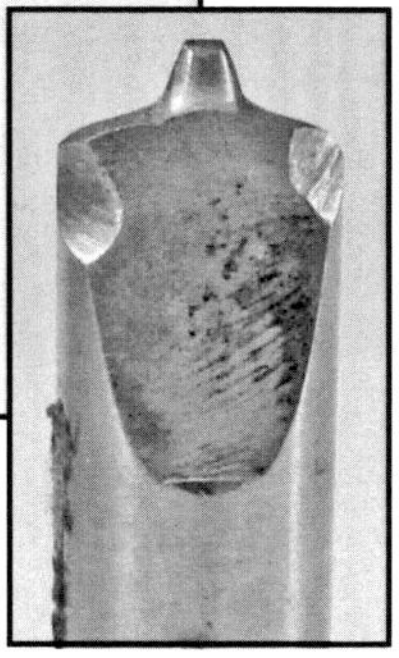

The inset photo (right) shows the two reliefs filed onto the back of the cutter, before it is hardened.

Making the Solid Pinions

There are three solid pinions in the movement. Let's take the cannon and minute pinions first. These will be made with the fly cutter which is shown in Figure 105. Make this fly cutter, using the procedures covered in Chapter 5 and Appendix B. It will cut satisfactory pinion leaves for the two pinions, even though they differ slightly in module. Figure 107 gives the dimensions of the cannon and minute pinions.

Fig. 106. A solid pinion in the chuck.

Figure 106 shows the appearance of a fly-cut brass pinion in the lathe. The desired lengths would be parted off from the piece.

The other solid pinion is the center pinion (Figure 108). It is really a wheel, but with only 32 teeth it has been included in this section on fly-cut pinions.

Cut the center pinion from a 1/8" thick brass blank. If you do not have 1/8" thick brass on hand, cut out two blanks from 1/16" clockmaking brass. Cement the two blanks together, then turn them to final diameter to create a single blank 1/8" thick.

Next, cut the teeth with the .82 module fly cutter. Figure 108 gives the dimensions of the center pinion. This is the gear which transmits rotation from the second wheel to the center arbor. The movement features the auxiliary drive method, which is common in American movements. In comparison, European clocks usually have the center arbor geared directly into the time train. In our American movement, the center arbor could be left out and there would be no gaps in the gearing from the main wheel up to the escape wheel.

The Arbors

Figure 109 shows the wheels and arbors for the clock. Four arbors are required for the second, third, fourth, and minute wheels. A longer arbor is required

...text continued on page 47

FIG. 107. CANNON & MINUTE PINIONS

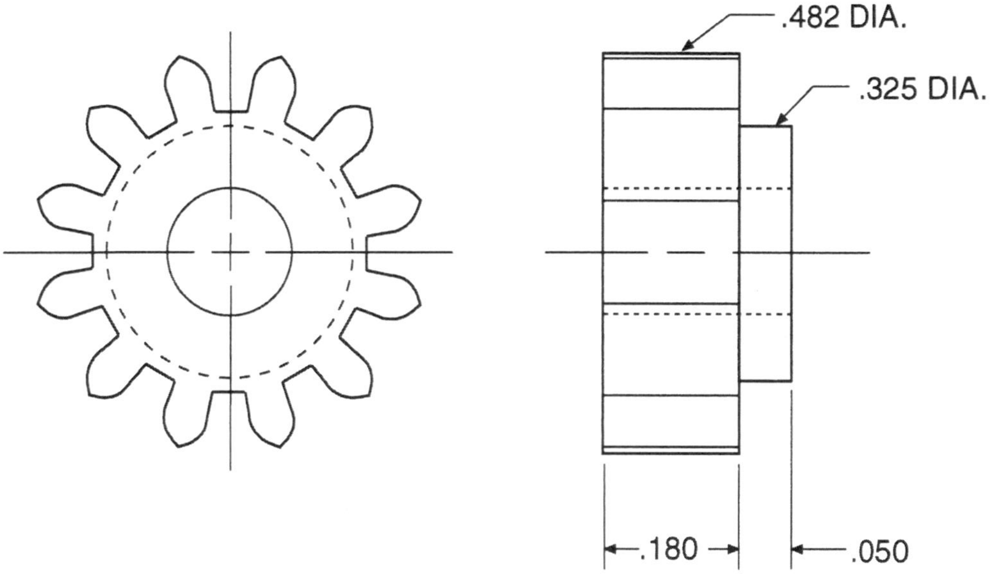

CANNON PINION
BEGIN WITH 1/2" BRASS ROD 2" LONG.
DRILL #13 AND REAM WITH ARBOR MATERIAL, 3/16" DRILL ROD.
CUT 12 LEAVES, TURN SHOULDER, PART OFF.

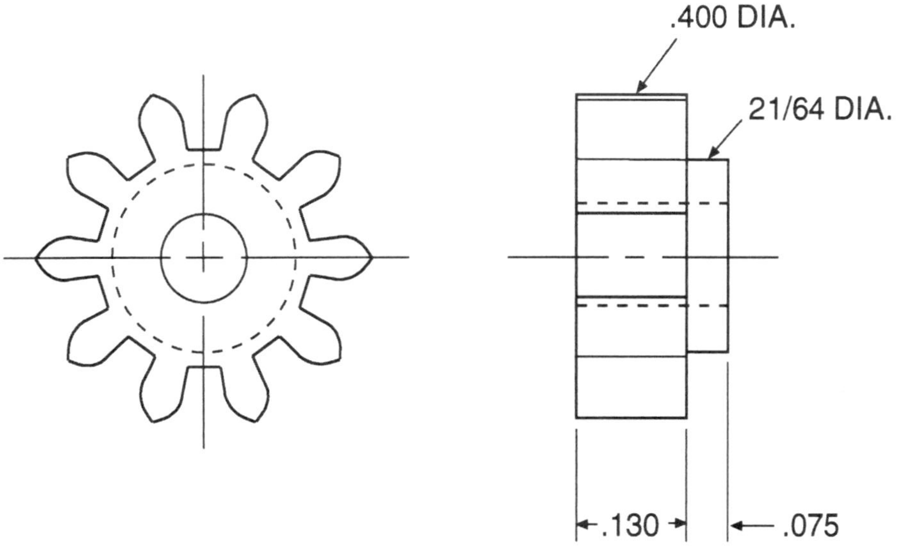

MINUTE PINION
BEGIN WITH 1/2" DIA. BRASS ROD 2" LONG.
DRILL #43, THEN REAM WITH ARBOR MATERIAL, 3/32" MUSIC WIRE.
CUT 10 LEAVES, TURN SHOULDER, PART OFF.

FIG. 108. CENTER PINION

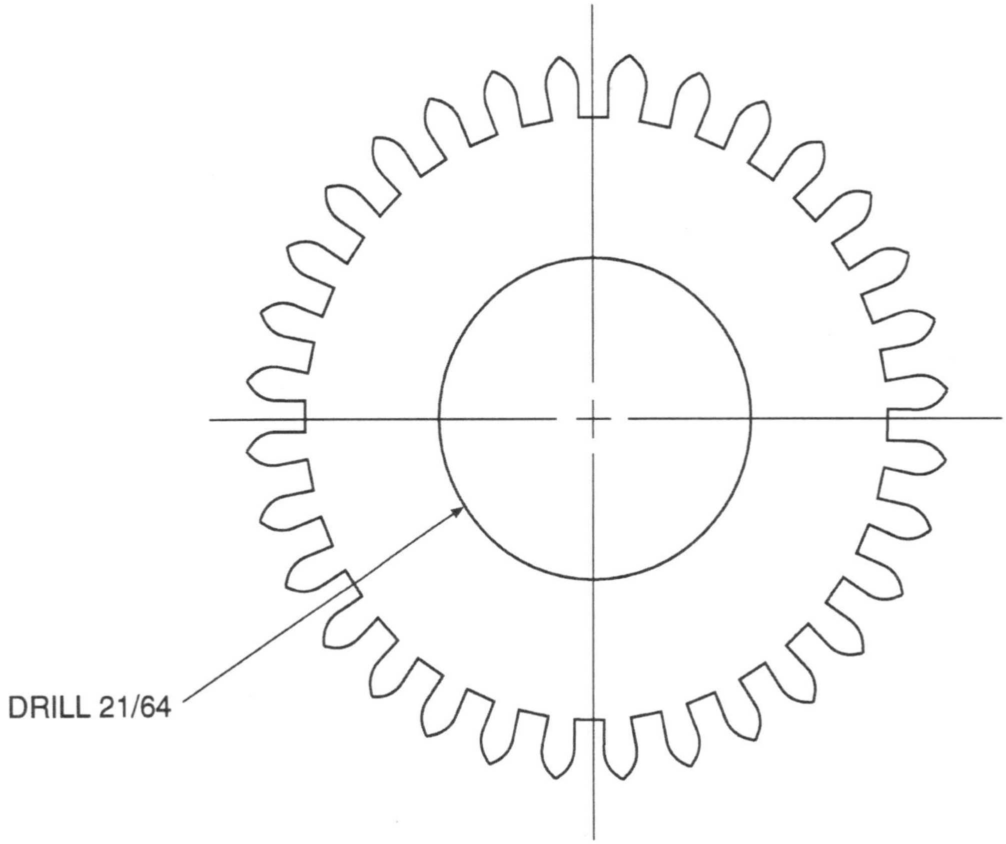

DIAMETER OF BLANK: 1.122". PITCH DIA. 1.033 ". ONE REQ'D. 32 TEETH
THIS PINION IS CUT AS A WHEEL USING THE .82 MODULE FLY CUTTER.
MATERIAL: C-353 BRASS SHEET, 1/8" THICK.
ALTERNATE: CEMENT TOGETHER TWO 1/16 THICK BLANKS BEFORE TURNING BLANK TO FINAL DIAMETER AND CUTTING THE TEETH.

for the escape wheel, and I suggest that this arbor be only partly finished now. It will be easier to finish it later, after the escape cock has been made.

The arbor dimensions are shown in Figure 110. The arbors are made from 3/32" diameter music wire. Although harder to work than drill rod, the music wire makes excellent arbors with tough pivots that will resist wear.

Four arbors. To make the four arbors for the second, third, fourth and minute wheels, mark the length shown on the drawing, Figure 110, plus about 1/8". It is not possible to saw music wire, so use the grinder to cut off the piece in the following manner John Wilding has described in his books. Apply the steel to the edge of the grinding wheel and twirl the steel slowly. Have a jar of water nearby to cool the piece. The result will soon become a conical joint between the arbor and the longer piece from which it is being separated. When the joint is thin enough, simply snap off the arbor. It is not necessary to grind it all the way through.

Fig. 109. The wheels and arbors.

Continue by using the grinder on both ends of the arbor until the ends are square and the total length of the arbor is correct at 1.595". Install the arbor in a watchmaker's lathe collet, with only a small length of arbor exposed. (Sherline and Unimat have adaptors which allow watchmaker's collets to be used. The point is to maintain concentricity, which cannot be done in the 3-jaw chuck.) Turn one end of the arbor to form a pivot of the correct diameter, .058". The diameter should not be any smaller, since the pivot hole will later be drilled in the plate. It is easy to slightly enlarge the hole in the plate if necessary, but an undersized pivot would make it necessary to install a bushing in the plate. Make the pivot .100" long. Lathe tool bits will cut music wire, but will become dull more quickly than when used on drill rod. Turn the arbor around in the lathe collet and turn the other pivot in the same way. The arbor should now have the correct length between shoulders. If the arbor is too long between shoulders, it will not fit between the plates. If it is too short, there will be too much endshake on the arbor.

Polish the pivots to a mirror bright finish. Begin by turning the arbor in the lathe and removing the tool marks with an abrasive paper such as a #400 grit. Use finer grits until a #2/0 grit brings up the polish. Much has been written about pivot polishing, which is certainly a subject unto itself. In making new pivots, however, we do not need to be concerned with the numerous problems encountered in repair work, such as deeply grooved, tapered pivots with rounded shoulders. If the pivot is formed with a lathe tool, it should only be necessary to polish it as described above.

It should not be necessary to polish the arbors. It is also important to avoid removing material from them, since the lantern pinions and wheel collets were designed for a slip fit, to be cemented later.

The escape arbor. Ron Pfleger made the suggestion that the escape arbor should be made after the escape cock, the reason being that it is more difficult to control the shape of the escape cock than it is to make the escape arbor a specific length between pivot shoulders. With this in mind, cut the rough escape arbor about 2-1/2" in length from a longer piece of 3/32" diameter music wire, using the method explained above. Turn one pivot only, making it .100" long and .052" in diameter. Place the escape arbor with the other four arbors for now; it will be finished later, possibly with a total length slightly different from that shown on Figure 110.

The Wheel Collets

The collets, or wheel seats, for the third wheel and the escape wheel can be made now. These are drawn in Figure 111. The seating for the second, fourth, and minute wheels has already been completed as part of the pinions; the main wheel is seated on the ratchet wheel; and the hour wheel seat is yet to come. The wheel collets are easy to make in the 3-jaw chuck, with concentricity guaranteed by the fact that each collet is made without loosening the piece in the chuck. A slight undercut of the shoulder between the smaller diameter and the larger one is optional but will assure that the wheel will press flat on the collet. When the collets are finished, they can be slipped onto the arbors without cementing them. The wheels for the clock will begin to look like the group in Figure 109.

FIG. 110. THE ARBORS (EXCEPT CENTER AND WINDING ARBORS)

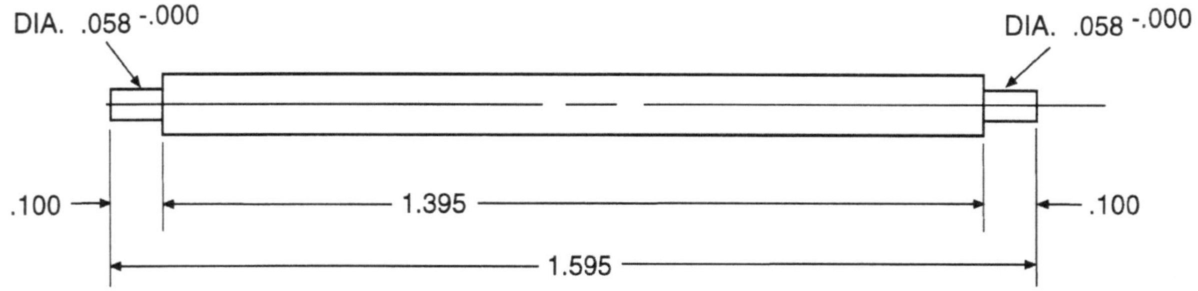

MAKE FOUR: 2ND, 3RD., 4TH AND MINUTE WHEEL ARBORS
MATERIAL: 3/32 DIA. MUSIC WIRE

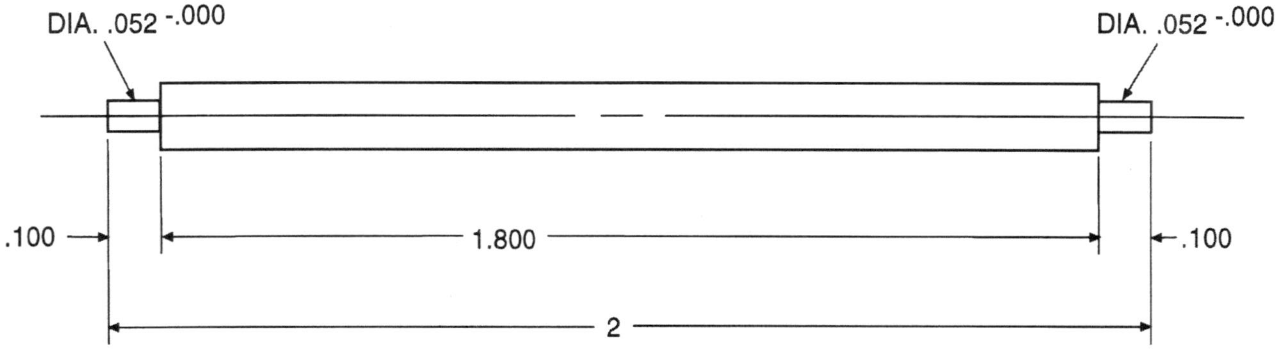

MAKE ONE FOR ESCAPE ARBOR
MATERIAL: 3/32 DIA. MUSIC WIRE

FIG. 111. THE WHEEL COLLETS

MATERIAL: 1/2" DIA. BRASS ROD

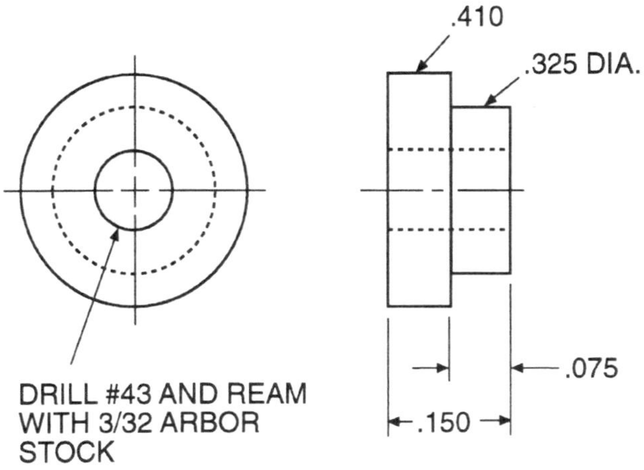

FOR THIRD WHEEL

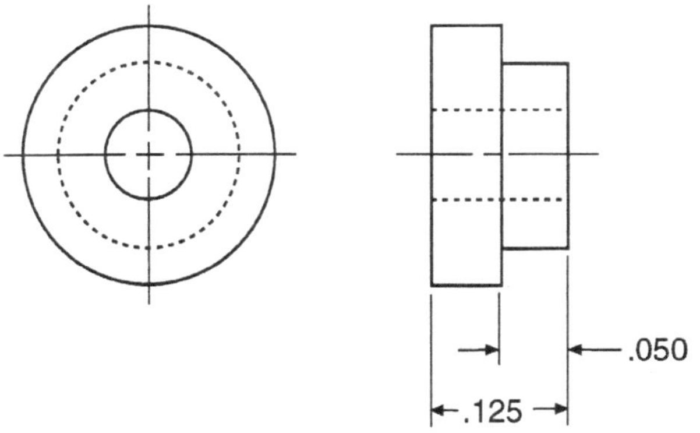

FOR ESCAPE WHEEL

SAME AS THIRD WHEEL COLLET EXCEPT AS NOTED

7
CENTER ARBOR ASSEMBLY

Fig. 112. The assembly without the hour wheel.

Fig. 113. A rear quarter view.

The center arbor is the basis of a main assembly consisting of the center arbor, hour pipe with hour wheel, cannon pinion, center pinion and hand clutch. Figures 112 and 113 are views of the assembly. The dimensions and fit of the finished parts are critical; just as the pivot shoulders are 1.395" apart for the second, third, fourth, and minute wheel arbors, the distance from the center arbor's rear pivot shoulder to the front of the hour wheel collet must also be 1.395". This will become clearer as the parts are constructed and fitted together.

Note that the hour wheel was described in Chapter 4 and the center pinion and cannon pinion were done in Chapter 6. The rest of the parts of the center arbor assembly must be made next.

Proceed to make the center arbor as shown in Figure 114. Directions in Chapter 5 covering the winding arbor will be helpful in making the square on the end of the center arbor for the minute hand. Do not drill the #55 hole which is located just to the right of center in the drawing. This hole takes the taper pin for the hand clutch spring, and it will be located by trying the fit later in this chapter.

Make the hand clutch spring from 3/4" diameter brass rod (Figure 115) and set it aside with the other parts of the subassembly. The curve of the spring

FIG. 114. CENTER ARBOR

MATERIAL: 3/16" DIA. DRILL ROD

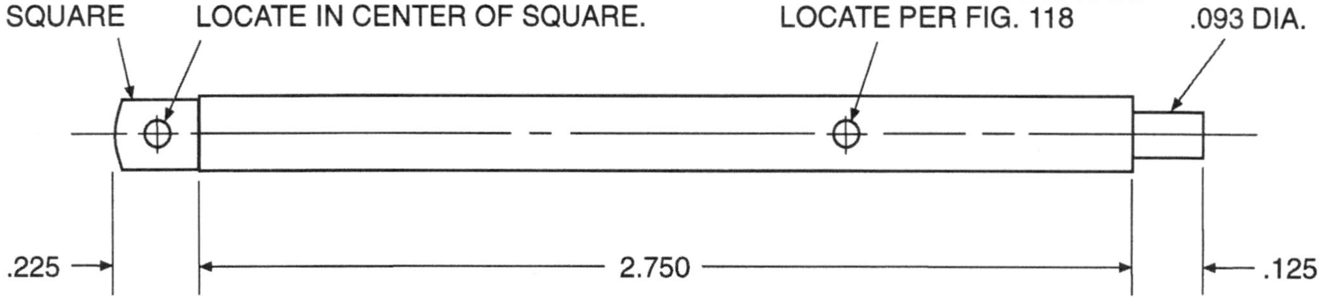

FIG. 115. HAND CLUTCH SPRING

MATERIAL: 3/4" DIA. BRASS ROD

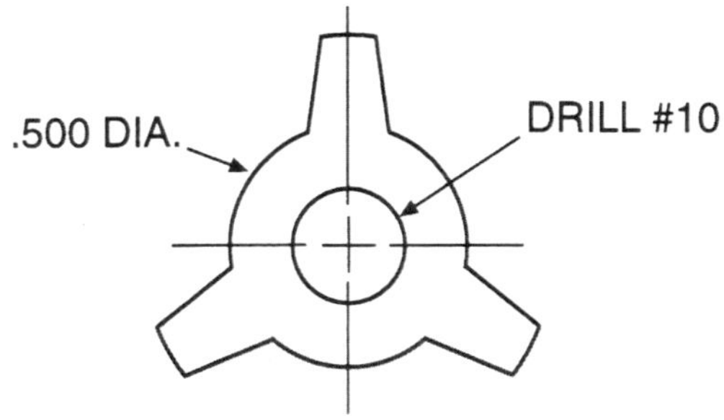

INSTALL PIECE IN 3-JAW CHUCK, DRILL #10, COAT END WITH LAYOUT DYE. SCRIBE 1/2" CIRCLE ON END. PART OFF .018" THICK. MARK OUT 3 LEGS OF SPRING, CUT OUT WITH JEWELER'S SAW, BEND LEGS TO SHAPE. SEE FIGURE 118.

can be made now or later, but it will have to be adjusted to give the correct hand tension. That step is covered later in this chapter.

Make the hour pipe assembly shown in Figure 116. Fit the parts together as shown in the drawing but do not cement the parts. Loctite 609 will be applied when the final assembly is being done.

Fitting the Assembly Together

The center arbor subassembly should be fitted together now. Based on fit and measurement, two items must still be located on the arbor: the cannon pinion, which will be cemented in place; and the #55 hole for the taper pin for the hand clutch spring.

Refer to Figure 117, which shows the complete center arbor assembly, except the hand clutch spring. The goal is to find the exact location to cement the cannon pinion onto the arbor. Place the dial calipers on the rear pivot of the center arbor and on the hour wheel collet. The distance between these two deter-

FIG. 116. HOUR PIPE ASSEMBLY

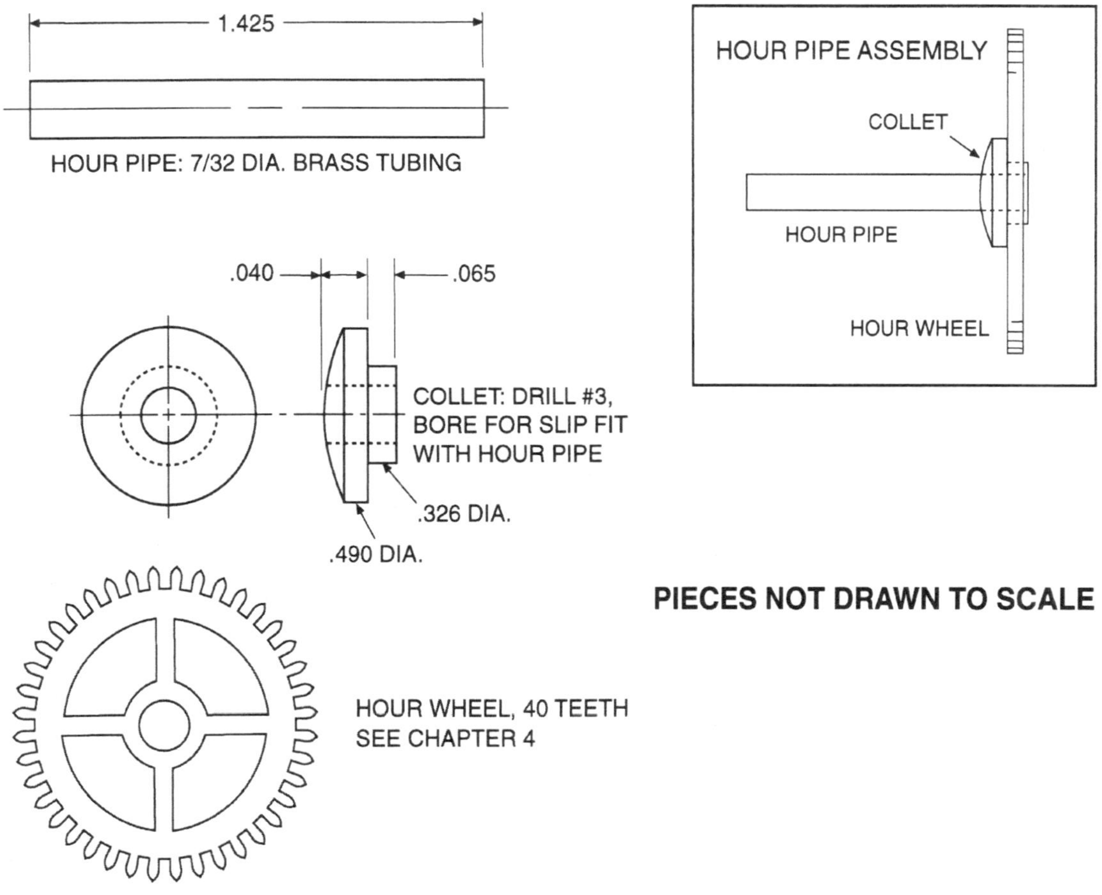

FIG. 117. LOCATING THE CANNON PINION

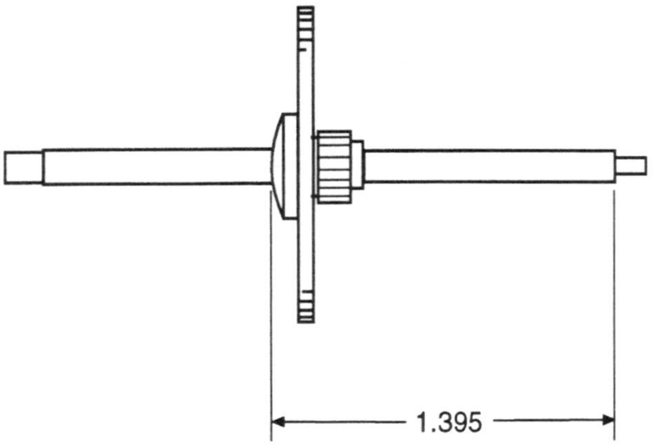

LOCATE THE CANNON PINION ON THE CENTER ARBOR BY ASSEMBLING AS ABOVE AND MEASURING 1.395" AS SHOWN. THIS IS A WORKING FIT FOR THE ASSEMBLY, WHICH FITS BETWEEN THE CLOCK PLATES (1.400" INSIDE CLEARANCE).
LOCTITE CANNON PINION ON CENTER ARBOR.

mines the endshake of the entire assembly. The endshake must be adequate (1.400" distance between the plates minus 1.395" equals .005") or the clock may not go together and work. If the parts fit freely on the center arbor, you can place the cannon pinion too far toward the front of the arbor, then watch the dial of the calipers as you bring the jaws closer together and drag the parts along. When the measurement reaches 1.395", stop. Carefully remove the hour pipe assembly, without moving the cannon pinion. Mark the location of the cannon pinion. Clean the surfaces with alcohol, then cement the cannon pinion in place on the center arbor with Loctite 609. Allow it to cure.

FIG. 118. LOCATING THE HOLE FOR TAPER PIN TO HOLD HAND CLUTCH SPRING.

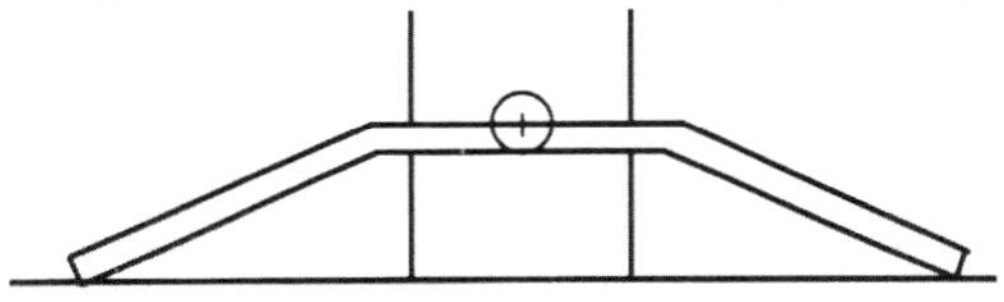

INSTALL THE CENTER PINION AND THE HAND CLUTCH SPRING. MARK THE HOLE LOCATION AS SHOWN. DRILL THE CENTER ARBOR #55 FOR TAPER PIN.

With the location of the cannon pinion established, slip the center pinion onto the arbor, which will then look like Figure 112. Now add the hand clutch spring behind the center pinion. Mark the location for the hole for the taper pin, as shown on Figure 118. Remove the hand clutch spring and center pinion from the arbor, then drill the arbor #55 at the marked location. When the assembly is put back together, select and fit a taper pin to the hole. It will be necessary to flex the spring slightly to allow the taper pin to go in. This provides the hand tension. Test the minute hand tension now by installing the minute hand or a substitute hand. Do not use pliers on the center arbor. The minute hand should move easily all the way around, but should not be so loose that it slips on its own. The tension should not be so great that it is difficult to move the hand. Adjust

Fig. 119. The center arbor assembly with hour wheel.

the minute hand tension by removing the taper pin, sliding off the hand tension spring, and bending the three legs slightly.

Place the hour pipe and hour wheel onto the front of the center arbor assembly, which will now look like Figure 119. A recheck of the measurement from the rear pivot shoulder to the front of the hour wheel collet should still show as 1.395". If the measurement is greater, you may need to reduce it, in order to have enough endshake for the assembly. One way to do that is to make a new hour wheel collet, which can be done because the hour pipe, hour wheel and wheel collet have not been cemented together. The best time to make the adjustment is later on, after the center arbor assembly has been fitted between the plates. On the other hand, if the endshake of the center arbor assembly is a few thousandths of an inch greater than the specified .005", there should not be a need to reduce it.

8
DEPTHING THE GEAR TRAIN

At this point in the construction, we have made the plates and pillars and a group of arbors with wheels and pinions. Now it is time to fit these arbors between the plates. As the holes are drilled in the plates, the parts will be fitted together many times. Figure 120 shows the movement at this stage.

Depthing the gear train means setting the distance between each of the arbors. The distance between one arbor and the next is equal to the center distance between the wheel on one arbor and its mating pinion on the next. In addition to setting the distance between arbors, it is necessary to design their arrangement. A gear train can, for example, be set in a vertical line from the main wheel up to the escape wheel, which is common in skeleton clocks. Or the gear train can curve generally around the center arbor, which reduces overall height. This is the case with our American movement.

Fig. 120. The partially completed movement.

The "correct" center distance can be arrived at through calculation, but for construction purposes it should be determined by installing the unmounted wheel and pinion on the flat depthing tool (the tool is pictured on page 3) and adjusting them for best fit. Trying both methods produced an interesting result. A center distance was first calculated, then the wheel and pinion were mounted on the flat depthing tool and meshed in the way that felt the smoothest as the pair was turned together. The two distances were consistently close to each other, with the distance on the depthing tool .005" to .007" greater than the calculated depth. It is better to use the settings from the flat depthing tool, not the calculated distances, to locate the holes for this movement. An exception is the distance from the center hole to the minute arbor hole, which is calculated.

Calculation Method

As mentioned in Chapter 4, gears mesh on their pitch circles (pitch diameters). This means that the distance between a meshing wheel and pinion is obtained by adding the pitch radius of the wheel to the pitch radius of the pinion. Pitch radius is simply number of teeth times module, with the result divided by 2. For a pinion of 8, we have 8 x .82 module = 6.56mm. This is divided by 2 to yield 3.28mm. Since we are working in inches, 3.28 is divided by 25.4 to give .129". The 60-tooth second wheel calculation is 60 x .82 = 49.2mm / 2 = 24.6mm / 25.4 = .969". To obtain the center distance between the 60-tooth second wheel and the eight-leaf pinion on the third arbor, add the pitch radii, .129" + .969" = 1.098".

As your reference, Figure 121 shows the calculated distances along with those established with the flat depthing tool. Also included is a table of the pivot hole sizes.

Drilling the Pivot Holes

Begin by coating the entire front plate with layout dye. Referring to Figure 121 for the measurements, mark the location of the winding arbor hole on the front plate. Punch mark this hole for later drilling.

Set the depthing tool to the distance from the winding arbor hole to the second arbor hole. To do this accurately, use the flat depthing tool. (A second choice would be the conventional type depthing tool.) Use the dial calipers to set the tool. This is easy to do as follows. Let's say your depthing tool has runners which are .188" in diameter. The distance between

FIG. 121. PIVOT HOLE LOCATIONS, SIZES, AND DISTANCES

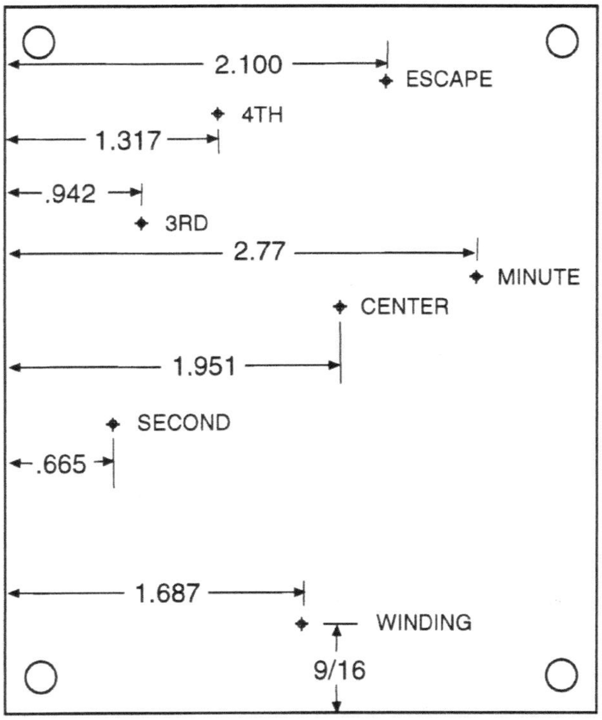

PIVOT HOLE LOCATIONS

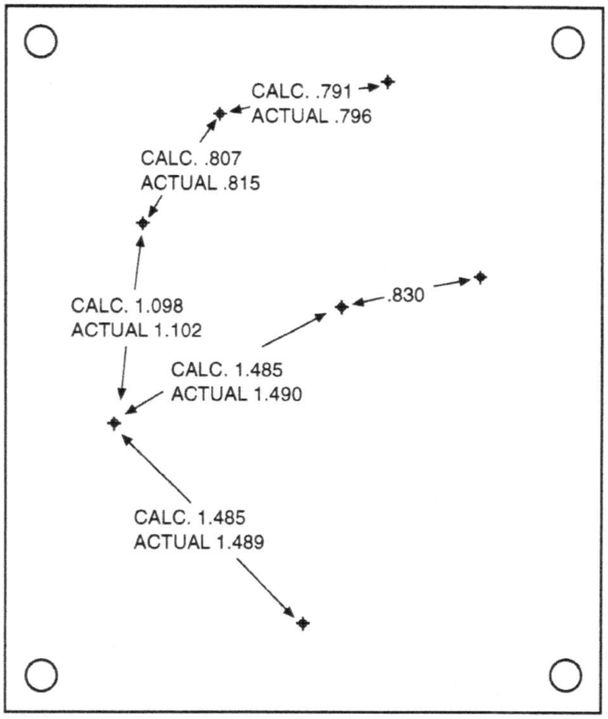

CENTER DISTANCES

PIVOT HOLE SIZES

WINDING ARBOR FRONT: DRILL 21/64 FOR BUSHING. SEE FIGURE 126
WINDING ARBOR REAR: SAME AS ABOVE
2ND: DRILL #53 AND OPEN WITH CLOCKMAKER'S BROACH AS NEEDED
3RD: SAME AS ABOVE
4TH: SAME AS ABOVE
MINUTE: SAME AS ABOVE
ESCAPE: DRILL #55
CENTER ARBOR FRONT (FOR HOUR PIPE): DRILL #2
CENTER ARBOR REAR: DRILL #42 AND BROACH TO FIT

CHECK YOUR DISTANCES WITH WHEEL & PINION IN DEPTHING TOOL BEFORE DRILLING

the points of the runners is the reading across the outside diameter of each runner, less the radius of each runner. To set the depthing tool with .188" diameter runners to a distance of 1.489", which is the center distance between the main wheel and the second arbor pinion, set the distance across the outer diameters of the runners at 1.677". This is 1.489" + .094" + .094". Since the runners are the same diameter, just add the full diameter of one runner. The result is the same.

With the depthing tool set to the center distance, one of the points of the tool can be set in the punch mark for the winding arbor hole, and an arc could be scribed onto the plate with dividers. The second arbor pivot hole will lie somewhere on this arc. But before scribing the arc, mark a point at the distance of the second arbor from the left edge of the plate, which is shown as .665". Use a square and rule a short vertical line at this point. Remember to scribe clearly but not so deep that extensive work will be needed later to polish out the lines. Now scribe the arc to intersect with the vertical line. Keep the arc short, about 1/4" in length. The intersection of the line and the arc is the location of the second arbor pivot hole. Punch mark it carefully for later drilling.

Proceed to scribe and punch mark the locations

FIG. 122. ESCAPE COCK

MATERIAL: 1/16" BRASS SHEET, ALLOY C-353

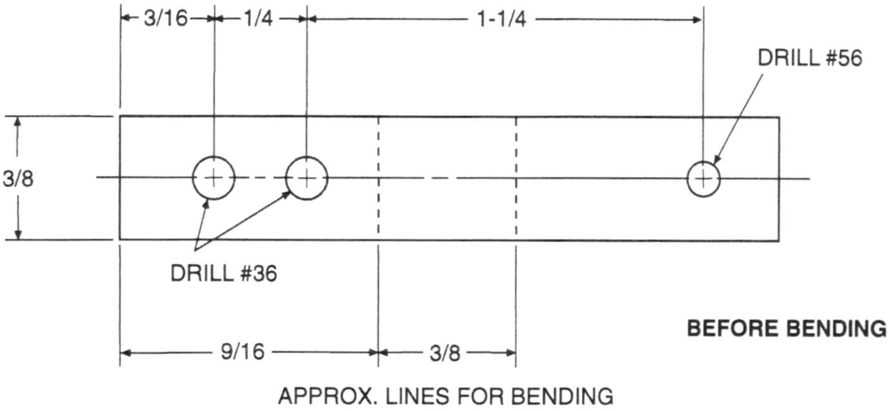

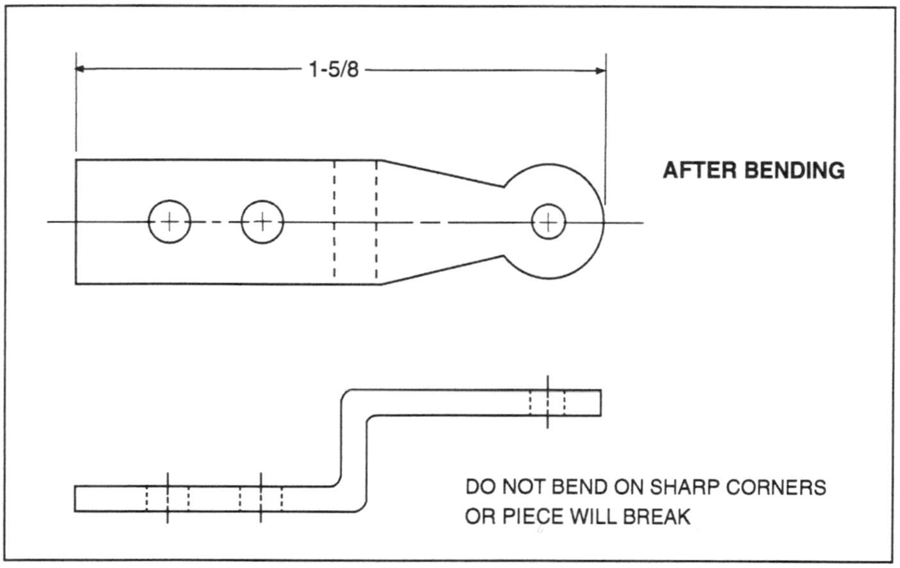

of the third, fourth and escape pivot holes onto the front plate. Each hole is "stepped off" from the previous one. The front pivot hole of the escape arbor will be carried on a cock, but we still need to mark the location on the front plate now, because it will help to place the escape cock accurately later on.

Put the front and rear plates flat together with the locating pins. Set up the drill press and drill a #53 hole through both plates at the second, third and fourth arbor locations. Then drill the escape pivot holes through both plates with a #55 drill. Next, drill the winding arbor holes through both plates with a 21/64" drill (bushings will be made and installed as bearings in the winding arbor holes).

Install the pillars in the rear plate with screws, then add the second, third and fourth arbors and pinions. Try installing the front plate to see which pivots fit in the holes and which will not fit, until the hole is broached out slightly with a five-sided cutting broach. You may want to fit the arbors one at a time. If an arbor is stiff, it may be because a pivot is too tight in the hole. There may also be a lack of endshake, and if that is the case the pivot shoulder must be cut back in the lathe to provide the endshake. Check the shoulder to shoulder distance with calipers, to see if it is close to 1.395". Since the three pinions and the third wheel collet have not been cemented, you can move them to allow the wheels and pinions to mesh. The pieces should hold well enough to allow the gears to rotate.

The Escape Cock

Separate the plates now and drill the front plate only, as follows. Drill the escape front hole out to #42. Broach the hole if needed to permit the escape *arbor* itself to pass through the front plate. This is a temporary move on our way to making and installing the escape cock. It is possible at this stage to drop the escape arbor through the front plate, slide the escape arbor pinion onto the arbor, and try the fit of the fourth wheel and the escape pinion.

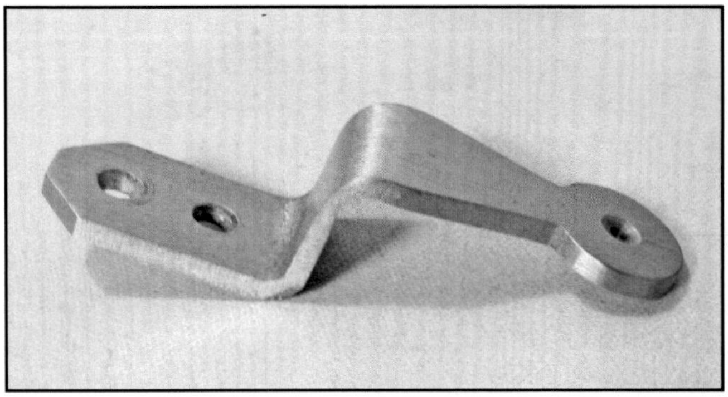

Fig. 123. The escape cock.

Figure 122 is a drawing of the escape cock. This can be made as three pieces soldered together, but I chose to bend it from a single piece of 1/16" clockmaking brass, which makes it resemble the same part on an old American movement. Try bending a scrap piece first. A piece can be put in the vise and bent over with a hammer, as long as the radius of the bend is not too sharp. I broke one piece before making a successful bend. Lay out a dye-coated piece of

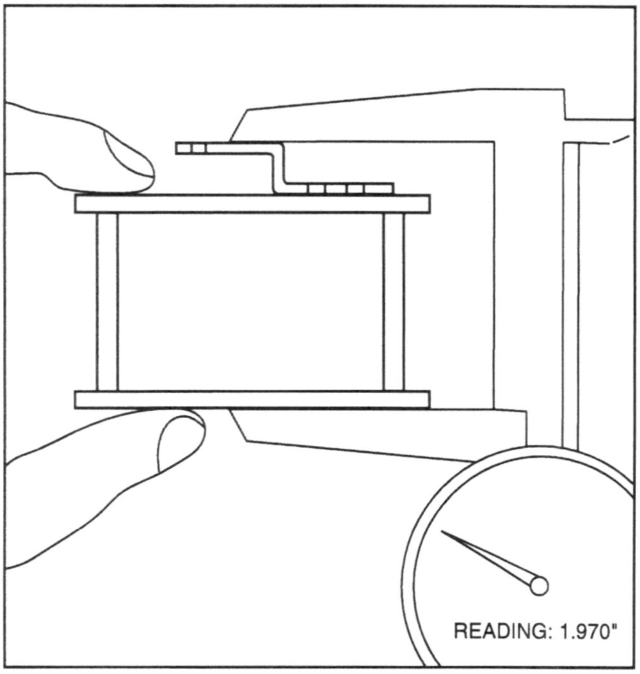

Fig. 124. Measuring for the length of the escape arbor.

brass with the outline, a center line, and the three hole locations. Punch mark the holes and either drill them now or bend the piece first and then drill. In either case, hold the piece in a machine vise for drilling. Do the finish work, including the shaping of the round section at the pivot hole location, after the successful bending work has been done. Figure 123 shows the finished escape cock.

Now determine the correct length for the escape arbor, which depends on the thickness of the movement and the dimensions of the escape cock, as it turned out after being bent to shape. Assemble the plates, hold the escape cock in place on the front plate, and place the calipers as drawn in Figure 124. The reading should be about 1.970". Subtract .130" from your reading (which is the thickness of the rear plate and the escape cock, plus some endshake) and you will have the length between shoulders of the escape arbor. The answer will be slightly different for every clock, because of differences in the escape cock; however, the distance between shoulders should be about 1.970" - .130" = 1.840". Shorten

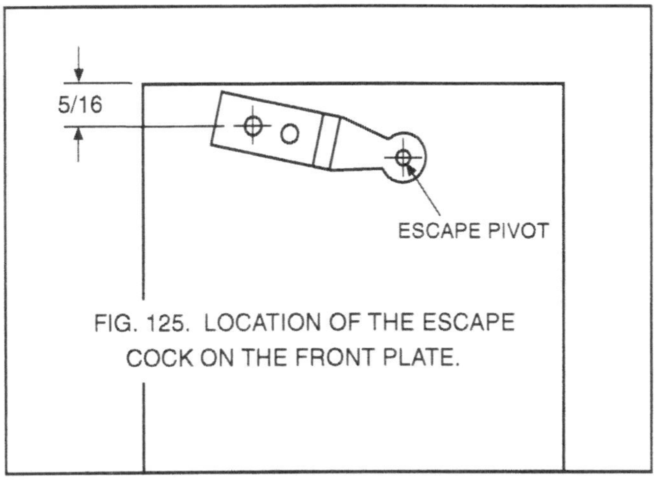

FIG. 125. LOCATION OF THE ESCAPE COCK ON THE FRONT PLATE.

the escape arbor as needed and cut the front escape arbor pivot. Again it should be emphasized that the length you establish for the escape arbor should be right for your movement. Note that on page 48 the drawing shows the length between shoulders as 1.800", but an answer of 1.840" was correct in the actual movement.

Fit the escape cock in place in the following way. With the plates assembled, drop the escape arbor through the hole in the front plate. Place the escape cock so the front escape pivot fits through the hole. Figure 125 shows the location of the escape cock. Mark the hole locations for the two screws which will hold the escape cock to the front plate.

Separate the plates again. Drill the two marked locations with a #43 tap drill. Tap the holes #3-48. Screw the escape cock to the front plate and then assemble the clock plates. The escape arbor should turn freely. Next, disassemble the plates and remove the escape cock. Drill out the temporary front escape hole, which was sized #42, to 25/64". This will allow the escape pinion body, which is 3/8" in diameter, to pass through the front plate from now on. The smaller hole served its purpose as a temporary bearing, a permanent function now performed by the escape cock. If the escape cock does not mount exactly as planned, or if the screws do not pass cleanly through the clearance holes in the escape cock to line up with the threaded holes, use a small round file to elongate one of the holes in the escape cock. There is no harm done if this correction must be made, unless the error is a large one.

Center Arbor and Minute Arbor Holes

Using the flat depthing tool and Figure 121, lay out the location on the front plate for the center arbor hole. Notice that it is stepped off from the second arbor. It is done this way because the second wheel is geared into the center pinion. Punch mark the hole. From this mark, lay out the location for the minute arbor.

Put the plates flat together with the locating pins. Drill the minute arbor pivot holes through both plates with a #53 drill. Next, drill the center arbor holes through both plates with a #42 drill.

Separate the plates again. Enlarge the front center arbor hole with a #2 drill. The resulting hole, .221" in diameter, will allow the hour pipe to pass through.

FIG. 126. WINDING ARBOR BUSHINGS

MATERIAL: 3/8" DIA. BRASS ROD

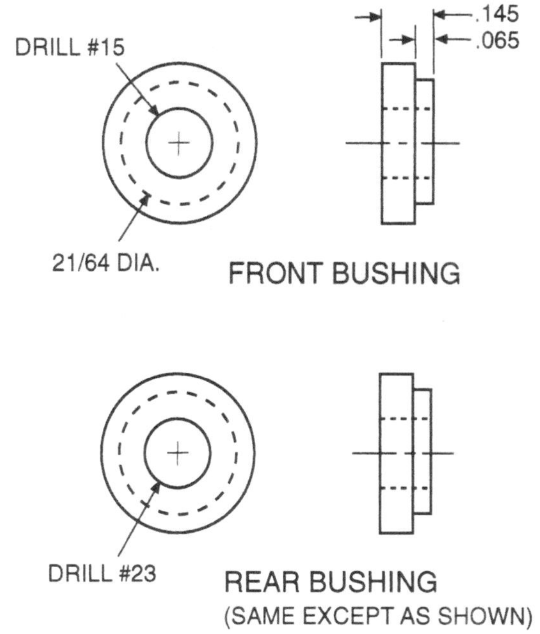

Winding Arbor Bushings

Figure 126 gives the dimensions of the front and rear winding arbor bushings. The bushings will press into the existing holes in the plates, and will be cemented or riveted permanently in place later.

9
THE ESCAPEMENT

Making an escapement can be a rewarding experience. After all, the escapement is the assembly which determines the accuracy and dependability of the movement. We made the escape wheel in Chapter 4. Now we are going to fit a set of strip pallets to the wheel. Purchased pallets costing only a few dollars will be used for the project and adjusted to fit. I did not want to complicate the project by making ordinary strip pallets.

Escapement terms can be confusing. The pallet unit is sometimes called the verge in the U.S., although that term properly has a different meaning, especially overseas. Also, the word "pallets" can denote the entire unit, even though the unit has two separate pallets on it, an entry pallet and an exit pallet.

Making the Pallet Cock & Fitting the Pallets

Begin by making the pallet cock, which is drawn in Figure 127. Have the pallet unit on hand so that the two will fit together. Specifically, the pallet unit has a bracket, called the saddle, which must slip onto the pin of the pallet cock with a fit that is free but not excessively loose. With the pallet unit in place, the pin on the pallet cock should extend about .030" outward from the saddle.

Figure 128 shows the finished dimensions of the pallet unit. As it comes from the supplier, the unit will have a long crutch wire. Shorten the wire, leaving more than enough to shape the wire to the dimensions shown. Figure 129 shows the crutch loop being formed with pliers.

The dimension of .721" (Figure 128) is shown as the span of the pallets. It will now be necessary to close (reduce) the pallet span as shown in Figure 130. If the pallet span needs to be increased to reach the .721" dimension, place two small steel rods across the vise jaws, then turn the pallet unit upside down across them. Rest the flat body of the strip pallet on the rods. Use the hammer and punch to tap on the underside of the saddle. This pallet opening and closing method was presented by James Tigner in

FIG. 127. PALLET COCK

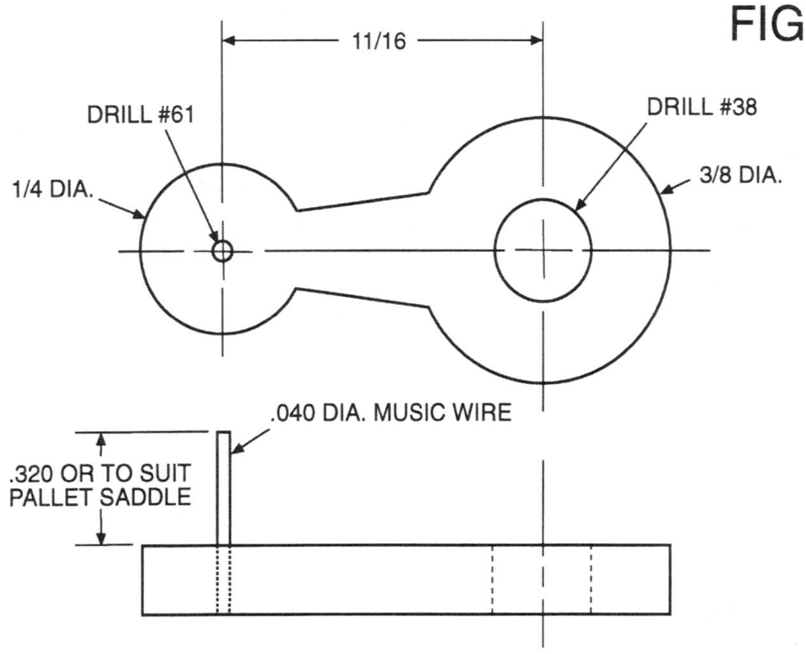

MATERIAL: 1/16" THICK BRASS SHEET, ALLOY C-353

FIG. 128. PALLET UNIT DIMENSIONS

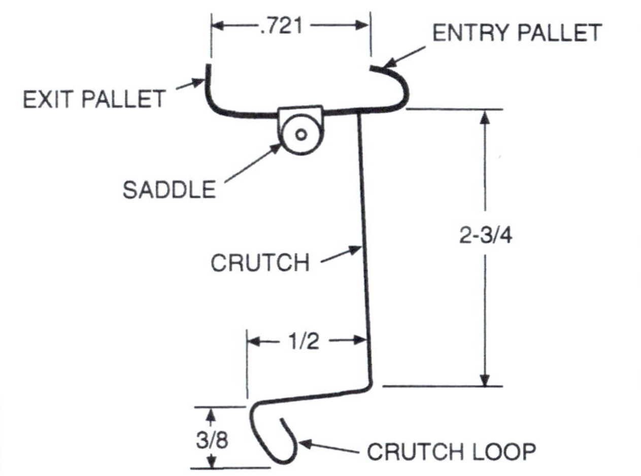

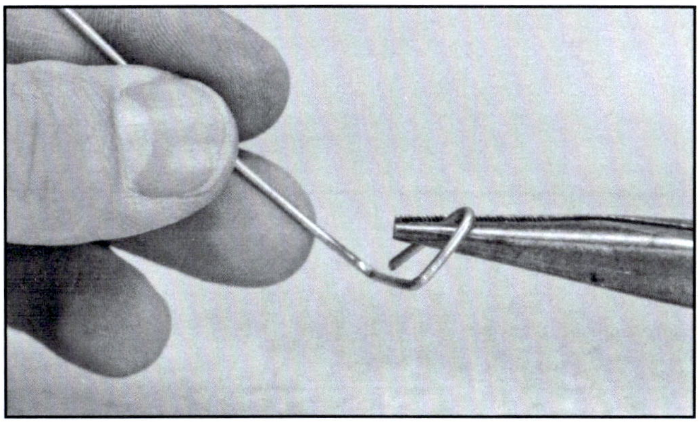

Fig. 129. Forming the crutch wire.

Fig. 130. Reducing the pallet span.

his writings for AWCI. Any attempt to bend individual pallets with pliers may break off the pallet or leave deep marks that will have to be filed and polished out. The dimension of .721" is the starting point for making final adjustments to the escapement, which will follow.

Figure 131 shows the escape wheel, still covered with layout dye, next to the pallet unit. The crutch wire has been formed and the unit set to the initial span as explained above.

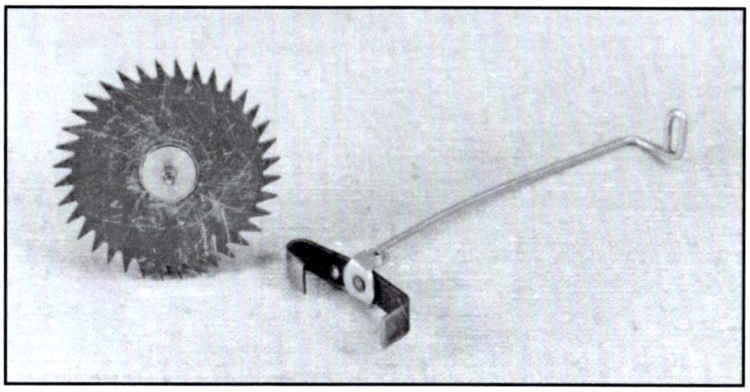

Fig. 131. The escape wheel and pallets.

Next, drill the hole for the pallet cock. The hole is located in Figure 132 and is a threaded hole in the front plate only. Screw the pallet cock to the front plate. As the escapement is adjusted, the screw will

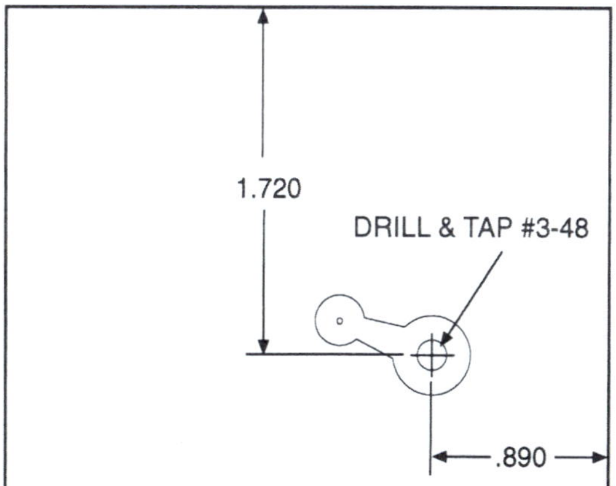

FIG. 132. LOCATION OF THREADED HOLE FOR PALLET COCK

be loosened and the pallet cock pivoted on the screw to bring the pallets closer to or further away from the escape wheel.

The Suspension Post

The suspension post should be made and fitted to the front plate at this point, so that the pendu-

FIG. 133. SUSPENSION POST

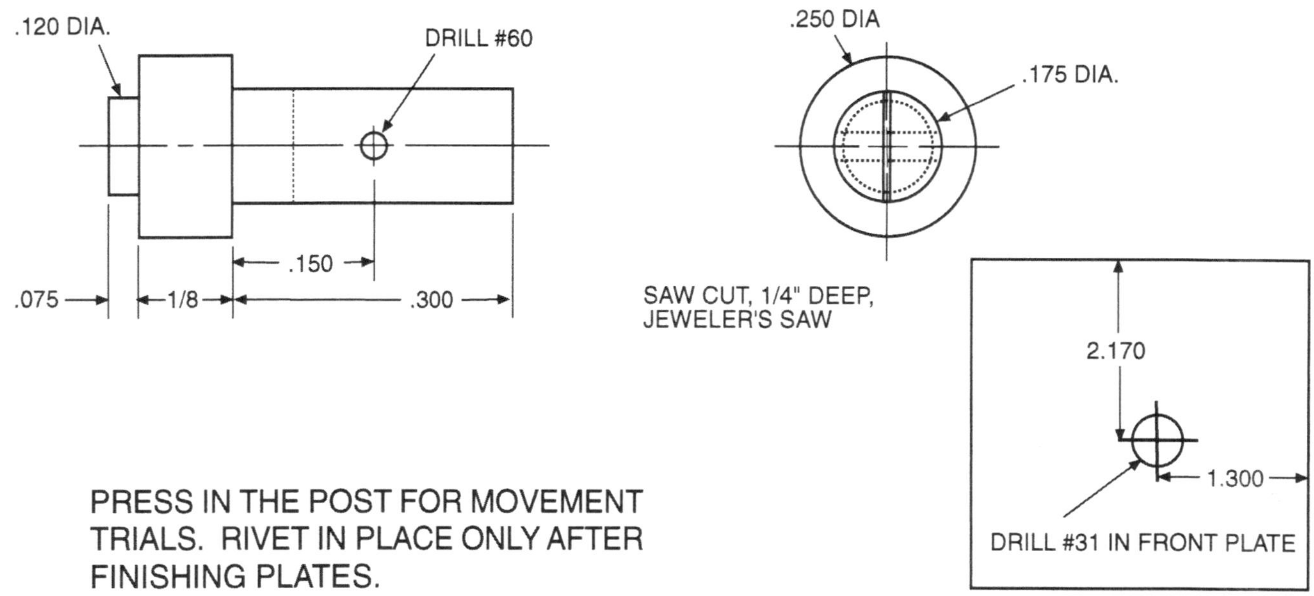

lum rod can be installed for testing the escapement. Figure 133 shows the dimensions of the suspension post and the location of the #31 drilled hole in the front plate. The post can be pressed into the hole and later removed so the front plate can be finished. Only at the end of the construction will the post be riveted in place.

Assembling & Adjusting the Escapement

Assemble the movement with the third and fourth wheels between the plates. After adding the front pillar nuts to hold the plates together, slip the escape arbor and pinion through the hole in the front plate. Mesh the escape pinion with the fourth wheel. Add the escape cock to the front plate and fasten it with two screws. The pinions may slide on the arbors, since they have still not been cemented together. This looseness may seem an annoyance now, but it is better to have the flexibility to move pinions and wheel collets until near the end of the construction, when all adjustments are complete. Install the pallet unit on the pallet cock. Set the initial position of the pallet cock so that the pin is .960" away from the escape pivot, as measured with calipers. This is not a final adjustment, but it should get the escapement functional. Install the pendulum rod in the suspension post and fasten it with a taper pin. The movement should now look like Figure 134.

Place the movement on a test stand or near the edge of the bench (carefully). Apply finger pressure to turn the gears so that the escape wheel turns clockwise, viewed from the front. Move the pendulum rod from side to side to make the escapement operate.

Fig. 134. The movement with escapement in place.

Loosen the screw on the pallet cock and turn the cock to move the pallets closer to the wheel or further away, until the escapement functions.

We have not yet made and installed the pallet retainer (Figure 136), so make sure the pallets do not work their way off the pin on the pallet cock. This would allow the escape wheel to spin against the pallet faces, which could damage the wheel teeth.

In Figure 134, the curved pallet on the right is the entry pallet; as the wheel turns, this pallet contacts the escape wheel teeth first. The straight pallet on the left is the exit pallet; the teeth touch this pallet second, on their way out of the pallet span.

Observe the teeth as they drop onto the two pallets. The amount of drop should be equal onto each pallet and close to the minimum which will allow the escapement to operate. If the drop of a tooth onto the straight exit pallet is too much, remove the pallets from the clock. Close the span of the pallets very slightly, by a few thousandths of an inch, using the method discussed in setting the initial span of the pallets. Try the pallets in the clock to verify that the drop has been reduced.

Now observe the drop of a tooth onto the curved entry pallet. To reduce this drop and make it equal to the other drop, bring the pallets closer to the wheel. This is easily done by loosening the screw and moving the pallet cock. Verify that the drops are now minimal and equal.

Once the pallets operate correctly, polish them with fine emery paper, using strokes that run with the length of the pallet faces, not across them. This places the strokes in the same direction the escape wheel teeth move as they slide across the pallet faces. Polish to a mirror finish, removing all marks and avoiding the rounding of edges. After initial polishing, the pallet faces should be hardened, one at a time. Heat a pallet slightly, then dip it in borax (Figure 135) or powdered boric acid, which prevents the formation of black scale during the heating process which follows. Now heat just the pallet, not the entire pallet unit, with the propane torch until it is red in color. Quench immediately in water. Check the pallet face with an old file. If the face is hard, the file teeth will not cut. Reharden if necessary. When the pallet face is hard, proceed to harden the other pallet face in the same way. Now polish both pallets again to bring up the mirror finish.

Fig. 135. Dipping the pallets in borax.

Pallet Retainer

Figure 136 shows the dimensions for the pallet retainer. A location is given for the mounting hole, which is made in the front plate only. Although the pallet retainer should be made up now, it is better left for installation in the front plate at the end of construction. If it is installed now, it just has to be removed when the front plate is polished.

FIG. 136. PALLET RETAINER

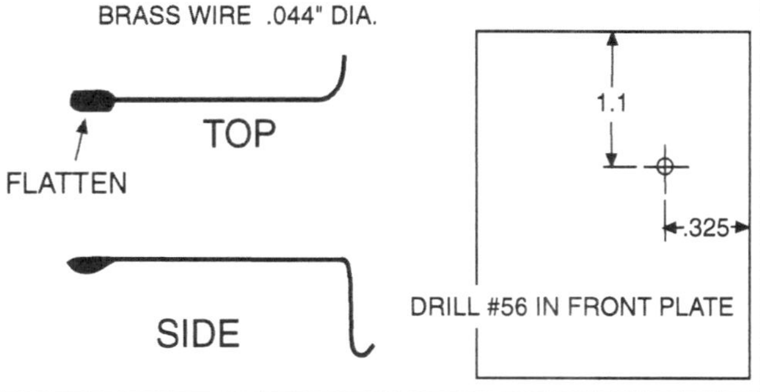

The suspension post can be pressed into the front plate, and it should hold well enough to support the pendulum rod and pendulum during a movement test. The suspension post is riveted permanently in place only at the end of the project because it, too, would get in the way of cutting out and polishing the front plate.

Figure 137 shows various escapement parts made so far, except for the escape wheel and pallets, shown in Figure 131.

...continued on the next page

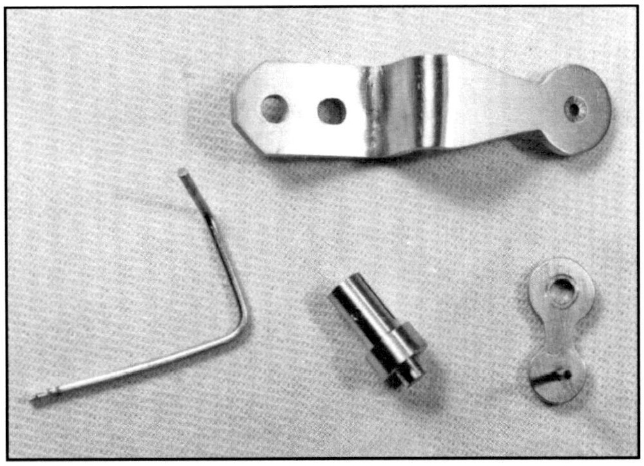

Fig. 137. Escapement parts, shown clockwise from the top: escape cock, pallet cock, suspension post, and pallet retainer.

FIG. 138. PENDULUM BOB

NOMINAL SIZE - MAKE TO SUIT CLOCK

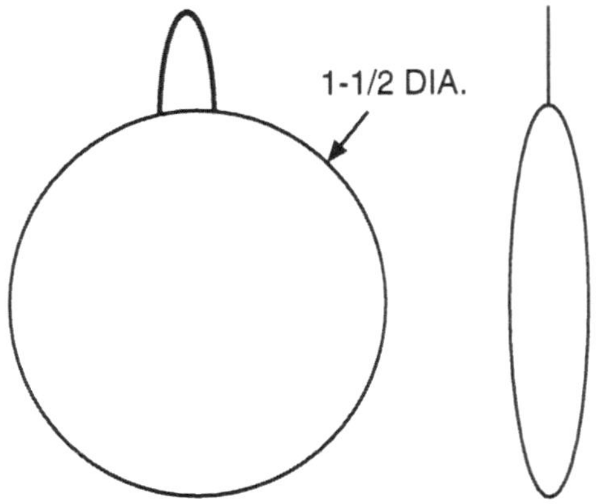

WEIGHT: 3 TO 5 OUNCES

The Pendulum Bob and Pendulum Rod

Figure 138 gives nominal dimensions and weight for a pendulum bob. Depending on the case you select for the clock, the bob may look different. Whatever the style, the clock is designed to beat 9072 times each hour. A compact pendulum bob such as the one shown in the illustration will work well, but a fancy kitchen clock style pendulum, which has decorative pieces attached to it, will also work. Some experiment is required. Be aware that a decorative pendulum, which has more weight higher up on the pendulum, must have a longer overall length, including the pendulum rod, than a simple round bob. Both can beat at the required 9072 rate, but the fancy pendulum will need slightly more case height.

The pendulum rod, which is available from clock parts suppliers, is cut to length and formed with a hook at the end for the pendulum bob. Depending on which bob is used, the pendulum rod length may differ. This, again, is a matter of experiment and is a part of designing and casing your own clock.

Thoughts on Making Changes to the Design

I will not advise readers to avoid making changes to the design described in this book. After all, a set of plans for building a clock mechanism naturally invites changes. Some readers have used the book as a starting place for making the movement they really wanted to make. However, be careful about those changes.

What about changing gear tooth counts to alter the pendulum length? You might later discover that some other part will not fit or work as desired, as a result of the change. Can you use different equipment? Other equipment might work much better or not at all in this application. A reader once asked about making the movement much larger. I could not predict the outcome.

A change in one area can bring about unwanted, unanticipated changes in seemingly unrelated parts of the movement. The best way to handle possible changes is to think through the entire design before going ahead.

In a related vein, builders have asked about changing the specified .82 wheel tooth module to .8 module, to permit the use of a purchased multi-tooth cutter in place of the home-built fly cutter. Changing the .82 module to .8 module will affect arbor-to-arbor distances only slightly. If you change the module size, be sure to work out how the wheel blank diameters and calculated distances between arbors will change.

10
FINISHING, ASSEMBLY & TESTING

More time may be spent on finishing and testing than on some of the main construction items, such as the actual wheel cutting. It is up to the builder to determine how much effort will be put into the finish. This chapter outlines the steps required to complete the project.

If you haven't already installed the mainspring, review the mainspring information on page 37. You must have the mainspring bound by a wire or a thicker steel loop. This provides a mainspring that is safely let down to a diameter that allows it to fit into the movement as you assemble it.

Fig. 139. Wheels and arbors in the rear plate.

The movement can now be put together for a short test run under a *small* amount of mainspring power. This can only be done if the wheels and pinions fit snugly enough on the arbors without cementing. The step of cementing will be done later, after final adjustments to the positions of wheel collets and lantern pinions. You will be winding the spring only enough to run the movement for a short time.

Figure 139 shows the appearance of the arbors placed in the rear plate, before the front plate is added. The escape arbor must be removed before the front plate goes on; the escape pinion then fits through the hole in the front plate. Adjust and mark the position of each pinion or wheel collet on its arbor. Make sure the second wheel contacts the center pinion. This ability to adjust the fit of the parts, before final attachment, is the main reason behind the temporary slip fit concept. Staked-on parts would be difficult or impossible to adjust.

At this stage, the assembled movement looks like Figure 140. The project screws still haven't been made—the plates and other items are held with ordinary steel socket head screws for now. Wind the mainspring up to ten "clicks" and there will probably be enough power to run the movement. Install the pendulum bob and start it swinging, making sure that the pallet retainer is in place. If the movement is ticking unevenly, adjust the position on the test stand or bend the crutch wire slightly one way or the other, until the ticking is even. Adjust the crutch wire so that the pendulum rod passes through the

Fig. 140. Movement assembled for its first test run.

center of the slot. The width of the slot should be adjusted to allow some clearance of the rod.

Run the movement for a few hours to confirm that it operates correctly. A list of possible causes for stopping would be a long one, but here are a few likely possibilities. Check each arbor for endshake, including the center arbor assembly. Also check each pivot for tightness in the hole. Recheck the escapement to determine whether it needs further adjustment. A pinion body might be rubbing on a wheel.

Seeing the movement run briefly under its own power for the first time is quite a thrill. The movement is far from finished, however. You can be brought down to earth in anticipation of disassembling the movement for cementing and finishing the parts. Many hours of work will be spent before the movement is put together again.

Finishing the Wheels and Arbors

Dimensions and procedures were given in Chapter 4 for crossing out the wheels. After the crossing out is done (it may take an hour or so for each of the wheels), each wheel can be given a straight line finish on a piece of Scotch-Brite abrasive. The wheels are not lacquered or otherwise finished, but all the scratches and traces of layout dye should be removed. Denatured alcohol removes the layout dye.

Fig. 141. Crossed out set of wheels.

Clean the parts with denatured alcohol, then cement the pinions and wheel collets to the arbors; cement the wheels to the wheel collets. Leave the parts to cure. Figure 141 shows the collection of finished wheels, arbors and pinions.

Cutting Out the Front Plate

The plates on the old American movements had sections stamped out to save brass. This could have been done to ease assembly or improve the appearance of the movement. Our movement has the front plate cut out for both reasons. It vastly improves the appearance of the movement. Besides giving it a finished look, the cut-out front plate allows the wheels to be seen much better than would be possible with a solid front plate.

Figure 142 gives dimensions and instructions for cutting out the front plate. A builder might decide, of course, to save time by not cutting out the plate. But anyone deciding to cut out the rear plate as well should bear an important point in mind: the rear plate is different. Specifically, it retains the rear escape pivot hole. For the front pivot hole, the escape cock provides the bearing.

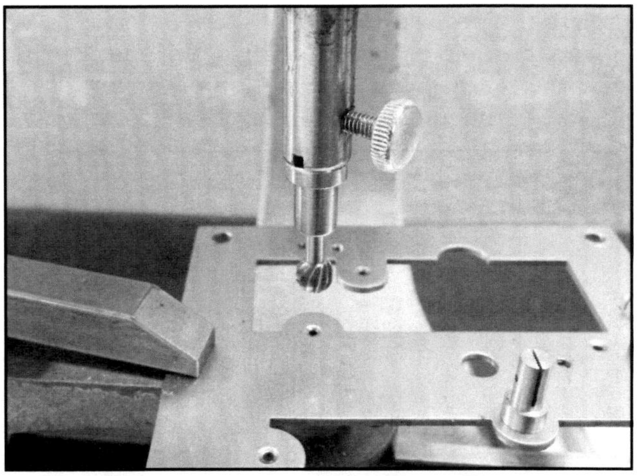

Fig. 143. Front plate in the bushing tool.

Figure 143 shows the cut-out plate installed in the bushing tool for the cutting of oil sinks. These could be countersunk by any means available in a shop, including a homemade D-bit, but the bushing tool is ideal for the task. A round burr is used to make oil sinks that are deep enough to allow the end of the pivot to just extend through the hole.

Use increasingly fine polishing papers on the plates until all the scratches are gone. Do not waste time trying to remove scratches with too fine a paper. A suggested finish is a straight line finish from Scotch-Brite abrasive. It gives an attractive look to the clockmaking brass.

Here are some suggestions for further finishing. The builder can clean the front and rear plates, suspension post, escape cock, and retaining wire (as separate parts), then spray on lacquer. As an alternate (which I used) wipe on clear lacquer which has been thinned 50% with lacquer thinner. It dries almost instantly unless a retarder, which is hard to get in small quantities, is mixed in. Nonetheless, good results are possible. A dust-free cloth must be used,

FIG. 142. SAWING OUT THE FRONT PLATE

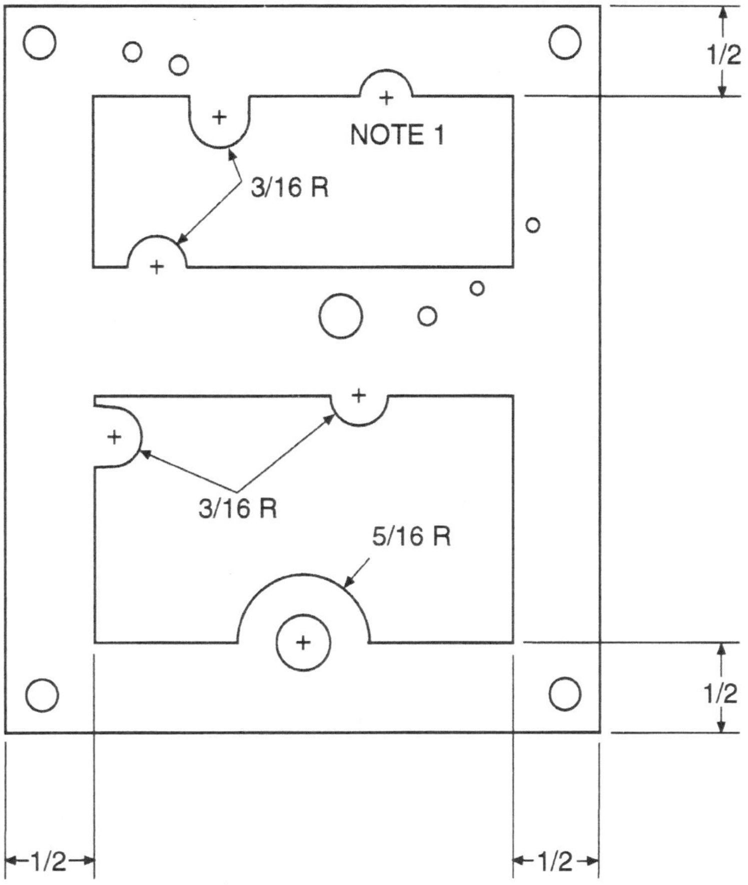

REMOVE ALL PARTS FROM FRONT PLATE.
COAT PLATE WITH LAYOUT DYE, AND LAY OUT LINES.
SAW OUT AND FILE TO LINES.
THIS APPLIES TO THE FRONT PLATE ONLY.

NOTE 1: THIS IS AN EXISTING HOLE CUT HALF AWAY. (HOLE TO ALLOW ESCAPE PINION TO PASS THROUGH FRONT PLATE.) IN THE REAR PLATE, THIS IS A PIVOT HOLE, NOT CUT AWAY.

and the pieces not handled until dry. And from then on, handle the parts with soft white gloves.

Final Assembly

Figure 144 shows the finished parts laid out before final assembly. Rivet the winding arbor bushings and the suspension post in the plates and add the pallet retainer. Load the arbors into the rear plate, as shown in Figure 145. The escape arbor is added later, through the front plate; the escape cock is removable, making it is easy to install the escape arbor and wheel after the plates are together. Finally, screw on the escape cock and add the pallets.

Screws

I have left the subject of screws until almost the end because it is up to the builder to decide whether to make or buy them. In the material section I have listed sources for screws. If you decide to make your own screws and blue them, follow the instructions here. Using drill rod, make the eleven screws specified in Figure 146, which can be cut in the lathe and threaded with dies.

The screw slots can be machine cut to produce the best possible appearance. Sherline and Unimat equipment can handle the task. A 1-3/4" diameter blade, .040" thick, was used. These were, in fact, the same blade and holder I used to make some of the edge-notched plates for the Huckabee Wheel Making Fixture. The .040" wide slots were good for the #4-40 screw heads, but this would have been too wide for the #3-48 screws.

For these smaller screws, I decided to use a 3/4" diameter blade which was only .018" thick. Since I did not have a holder for this blade, I decided to make one.

Fig. 144. Movement parts displayed on the bench.

Fig. 145. Crossed out wheels in rear plate.

Fig. 147. This saw blade holder consists of a threaded body, a circular blade, a cap, and a #4-40 socket head screw with washer. It was used for slotting the #3-48 screws in the movement.

Figure 147 shows the saw blade holder I made. Before making one, readers should note that Sherline makes a holder (see the Tools section in Chapter 1). The 7/16" diameter body of the homemade holder fits in the 3-jaw chuck. The threaded end is turned down to .312" diameter to accept the hole in the blade. Of course, a different saw blade holder would be required for a blade with a different size hole.

Figure 148 is a drawing of the homemade holder. The counterbore on the cap assures that the outer edge of the cap, rather than the center portion, presses on the blade for a secure holding force.

The overall setup for the slotting of screws in the Sherline mill is shown in Figure 149. The saw holder is gripped in the 3-jaw chuck, and the screw is secured in the Sherline milling vise. The saw is rotated in the headstock unit. The screw is moved past the saw with the cross feed handwheel, which faces the viewer in the photo. This is the "Y" axis of the mill. The table feed handwheel ("X" axis) is used initially to set the depth of the screw slot. The slot depth was set at half the height of the screw head.

Figure 150 shows how a screw was placed in one end of the milling vise, in the groove which holds

FIG. 148. SAW BLADE HOLDER

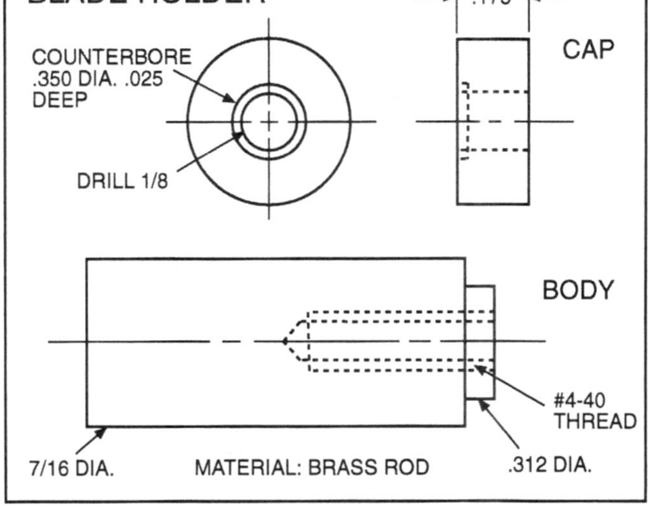

FIG. 146. SCREWS

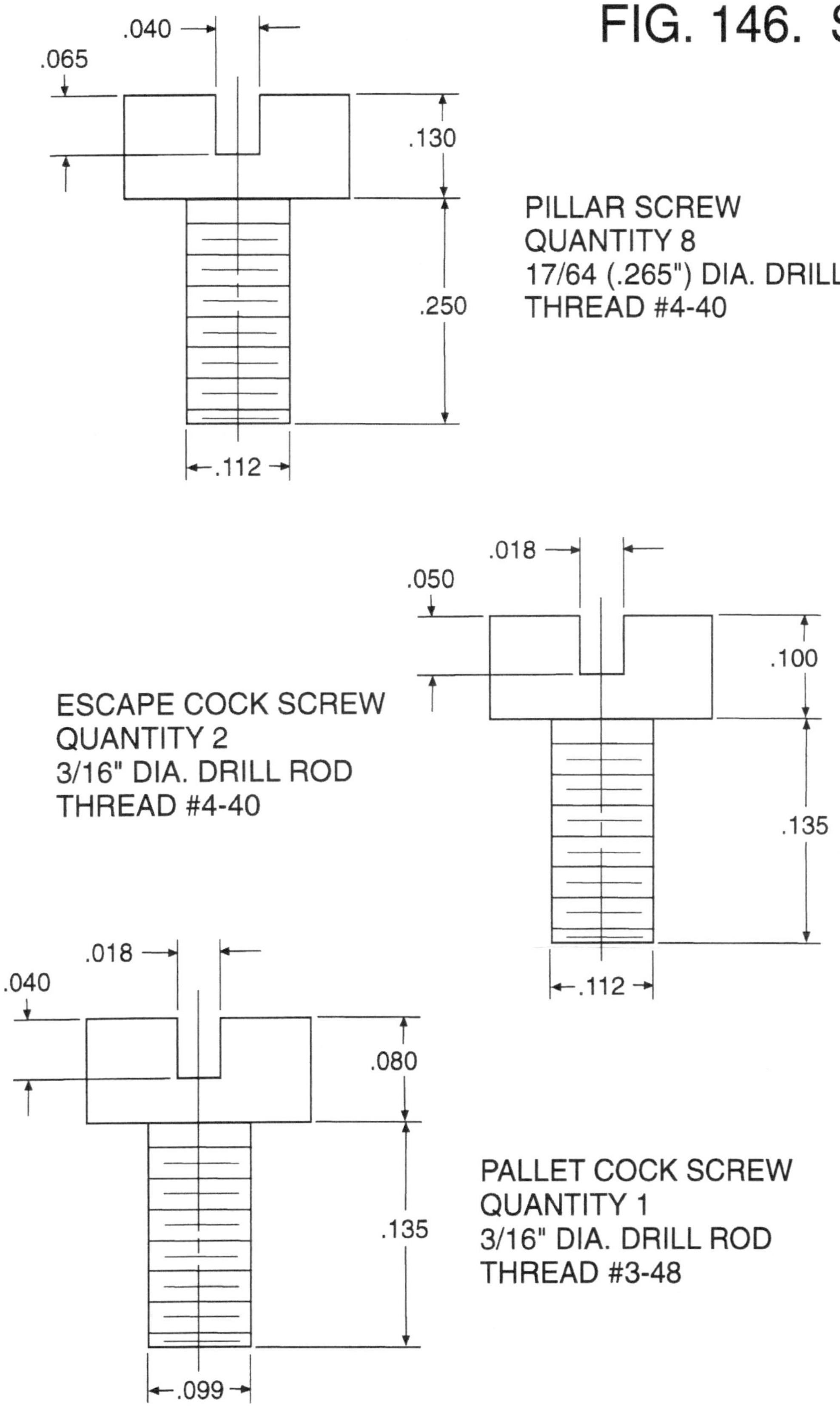

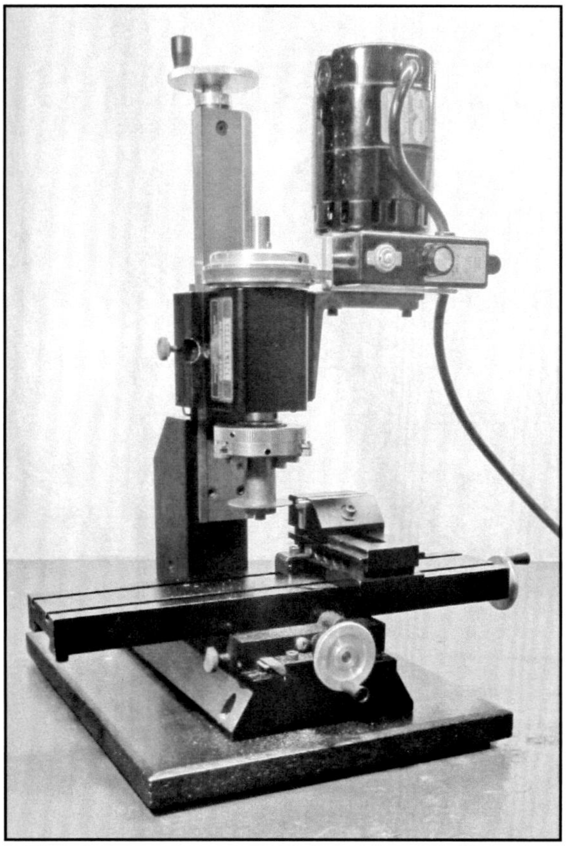

Fig. 149. Sherline set up for cutting screw slots.

round pieces. Another piece of the same diameter threaded steel was held at the other end of the vise to keep the jaws parallel. The handwheel was turned slowly to move the screw smoothly through the rotating saw blade. After the cross feed hand wheel was brought back to the starting point and the machine was switched off, another screw of the same size was placed in the vise for slotting.

Fig. 150. Cutting a screw slot in the Sherline mill.

After the slots were cut, the screws were heat treated to turn them blue. The blued screws greatly improve the appearance of the movement. Bluing is not difficult to do on screws. I used a watchmaker's lathe to polish the screws first. After burrs were removed from around the slots, I used several grits on the screw heads, finishing up with #2/0. The flat top of the screw was carefully drawn across abrasive papers on a bench top. When the finish was good, I placed the screw back in the lathe and held the #2/0 paper up to the end. It produced a good finish. To blue a polished screw, clean it with alcohol and do not touch it with the fingers after that. A small (3" x 3") piece of flat brass has a hole drilled in it to accept the screw. The brass is held over an alcohol burner

Fig. 151. The complete set of screws.

flame until the screw head turns straw yellow, then brown, then purple, and finally a deep blue. At this point, tip the screw out of the hole into a jar of water to stop the color changes. If the screw head turns light blue, it has gone too far and must be repolished and then blued again.

Figure 151 shows the full set of 11 blued screws. I think it is worth the effort for a handmade movement.

Completion of the Movement

The movement should be oiled with a good quality clock oil. Place a small drop in each pivot hole and on each pallet face. Also oil the saddle of the pallet unit. The mainspring, as mentioned earlier, was lubricated before assembly of the movement. Clock grease can be coated on the tip of the click or a ratchet tooth, and on the crutch loop.

The book has not included any drawings of mounting methods, since the builder will have his or her own ideas. Tabs can be made and screwed under each of the four rear (or front) pillar screws. This is the simplest idea and is similar to the method used in most shelf clocks.

Similarly, I have left hands out of the project, except for one detail—the hour hand bushing. This is shown in Figure 152. The minute hand, whether purchased or made from scratch, needs a square hole to fit the square on the end of the center arbor, a washer, and a taper pin. The hands will, of course, be designed to fit the individual dial size and style. Figures 153-155 conclude the project with three more photos of the movement.

Happy clockmaking!

FIG. 152. HOUR HAND BUSHING

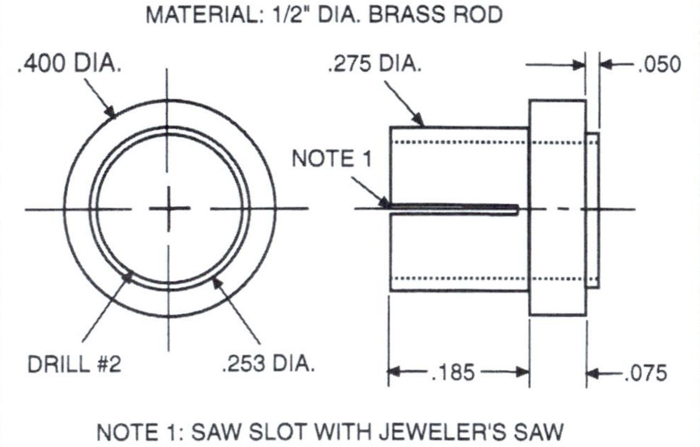

Fig. 153. Close-up through front plate.

Fig. 154. Rear quarter view.

Fig. 155. The finished movement.

APPENDIX A
THE HUCKABEE WHEEL MAKING FIXTURE

Reproduced with the kind permission of J.M. Huckabee
and the American Watchmakers-Clockmakers Institute

WHEEL MAKING FIXTURE

FOR MAKING CLOCK WHEELS

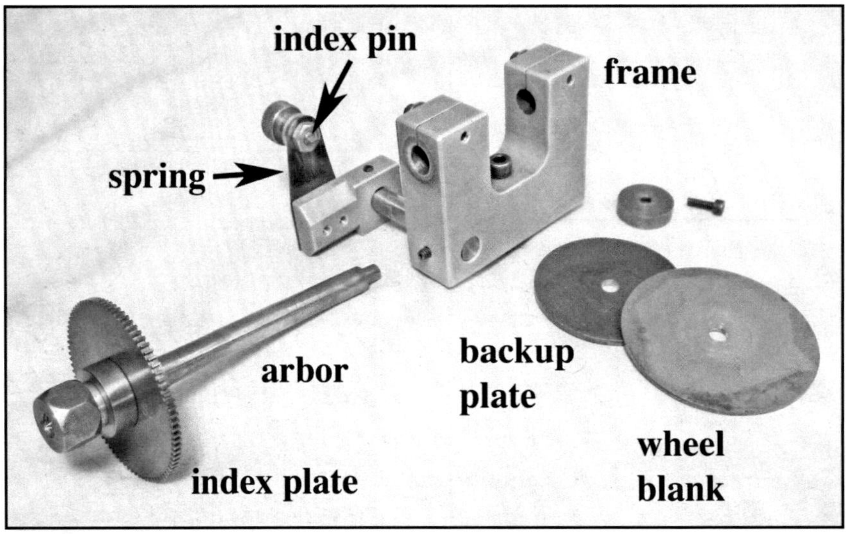

- DESIGNED BY J.M. HUCKABEE
- ILLUSTRATED IN RANDOM CLOCK TAPES 12 & 13
  (2- 2HOUR VHS VIDEO TAPES)

- FOR USE WITH THE UNIMAT LATHE, AND OTHER LATHES OF SIMILAR SIZE & TYPE

JUNE 1988

INDEX FIXTURE MAINFRAME

SHEET 1 OF 10

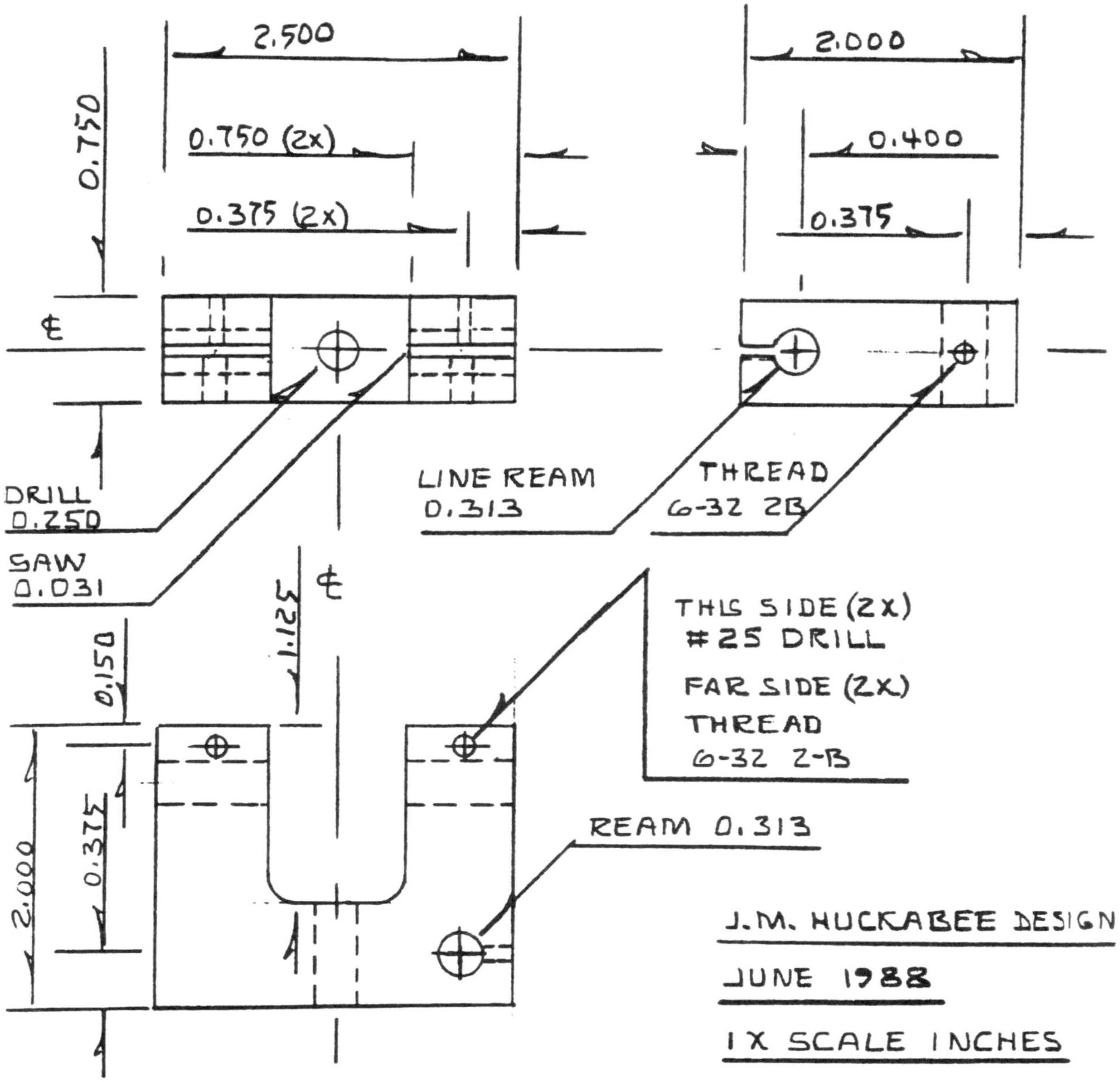

MATERIAL: HARD ALUMINUM

FINISH: BREAK ALL EDGES
WIRE-BRUSH TO SATIN FINISH.
WASH PROMPTLY TO PRODUCE A SELF-OXIDE PROTECTIVE COATING.
WASH WITH HOT WATER AND STRONG SOAP

SUGGESTION: DRILL & LINE-REAM 0.313 HOLES BEFORE MAKING 0.031 SAW SLOTS.

SHEET 2 OF 10

INDEX FIXTURE WORK ARBOR

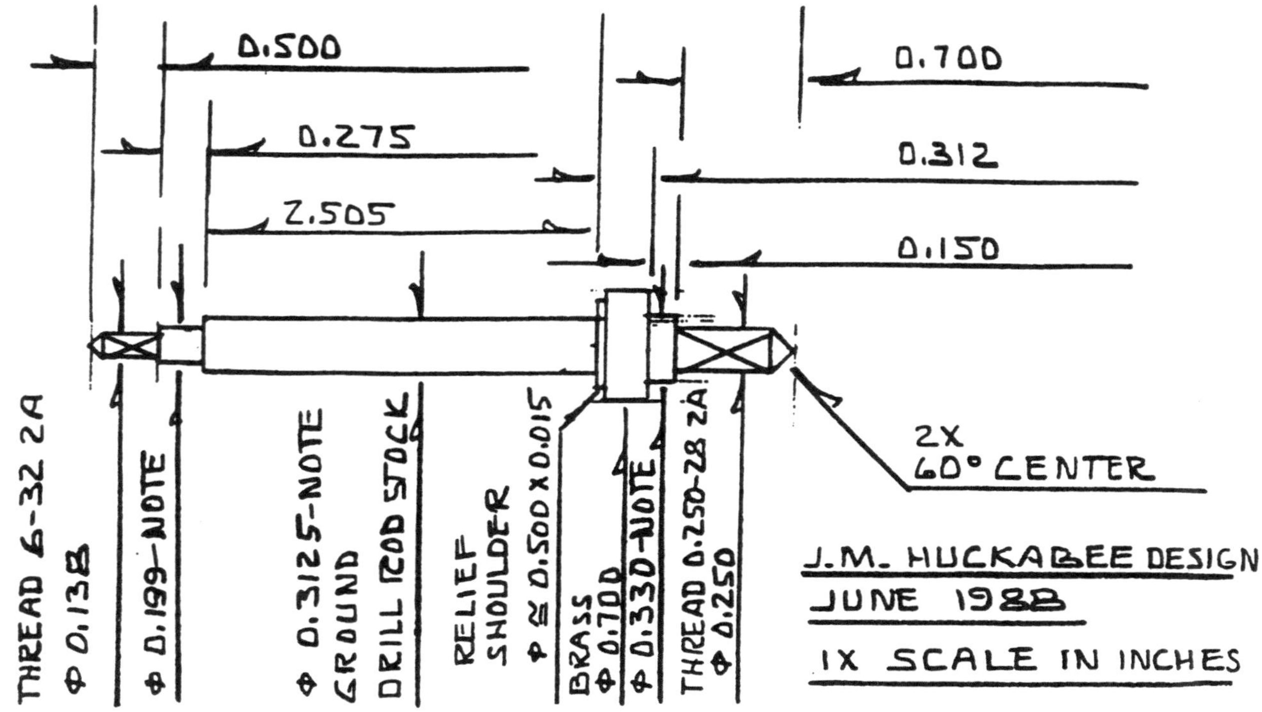

MATERIAL: DRILL ROD | BRASS

NOTE: THESE DIMENSIONS MUST BE CONCENTRIC

SUGGESTIONS: LATHE-TURN A 0.308 HOLE 0.275↧ IN THE END OF A PIECE OF 3/4 X 1-1/2 BRASS STOCK—SQUARE THE ENDS. SQUARE THE ENDS OF DRILL ROD STOCK ⊈ Δ 0.005. PRESS-FIT WITH ≅ 800–1200 # FORCE.

HOLD TIP OF DRILL ROD IN A TRUE COLLET, SUPPORT WITH STEADY-REST AND FINISH BRASS SECTION.

NOW SUPPORT UNTHREADED BRASS SECTION IN A TRUE COLLET, USE STEADY REST, AND FINISH DRILL-ROD SECTION

SHEET 3 OF 10

INDEX FIXTURE-SPRING BLOCK

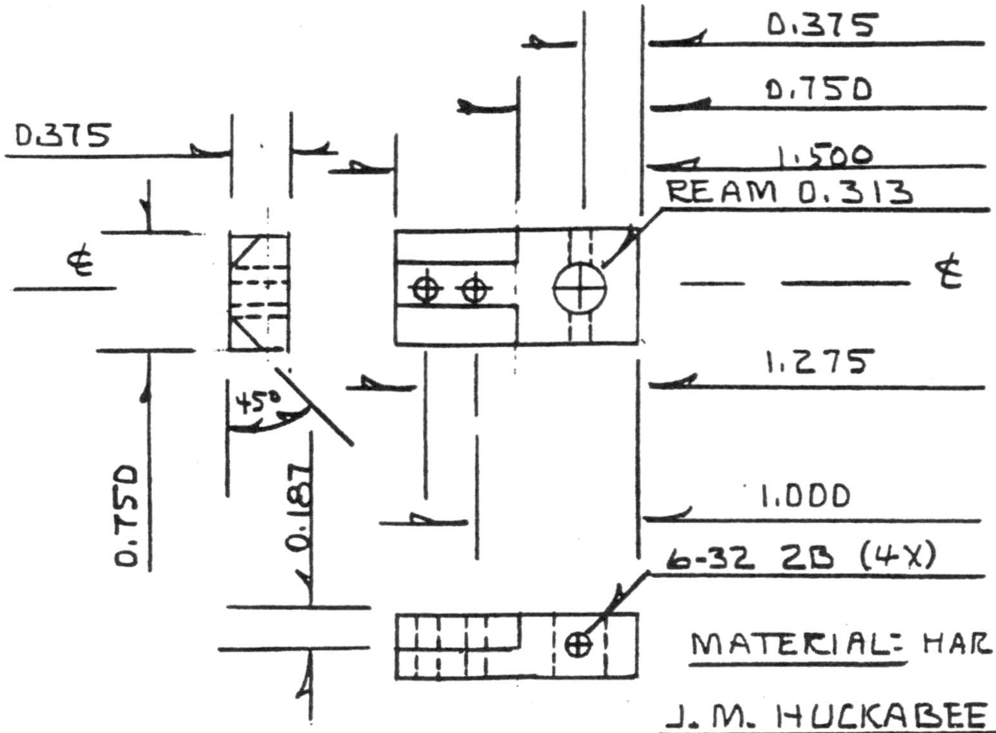

MATERIAL: HARD ALUMINUM

J. M. HUCKABEE DESIGN
JUNE 1988
1X SCALE IN INCHES

INDEX FIXTURE-SPRING

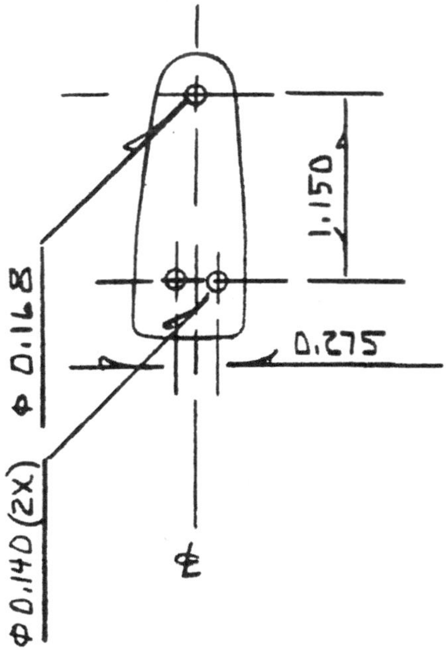

MATERIAL: CLOCK MAINSPRING APPROX 0.750 WIDE. 0.017 THICK.

SUGGESTION:

GRIND TO APPROXIMATE SHAPE DEPICTED. DO NOT REMOVE CURVED SHAPE, PIERCE OR GRIND HOLES - PRESERVE TEMPER

SHEET 4 OF 10

INDEX FIXTURE-SUPPORT ROD

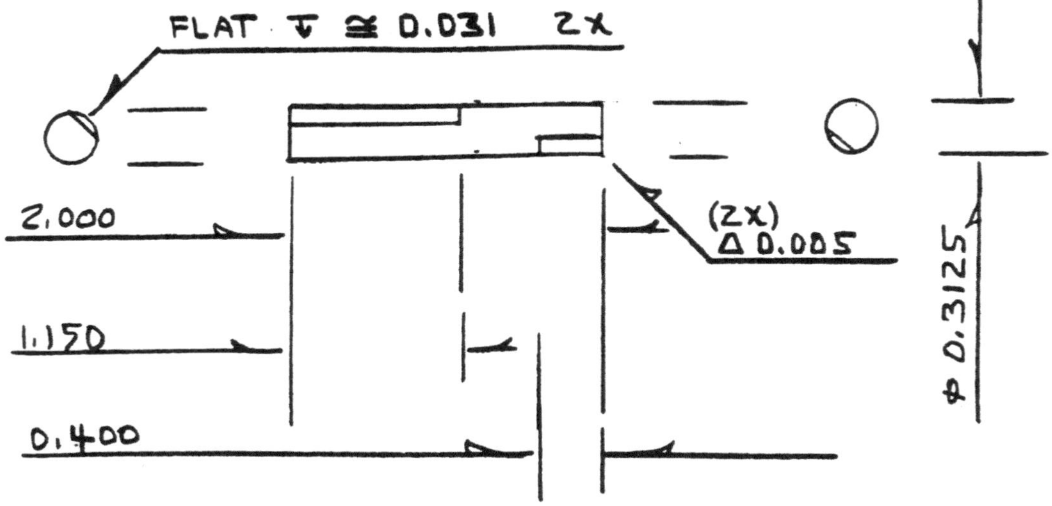

MATERIAL: GROUND DRILL ROD STOCK

J.M. HUCKABEE
DESIGN

JUNE 1988

1X SCALE - INCHES

INDEX FIXTURE NUT & SPACER

MATERIAL: 9/16 HEX BRASS STOCK

SUGGESTION: CUT SHAPE APPROX AS DEPICTED. THREAD 1/4-28 2B

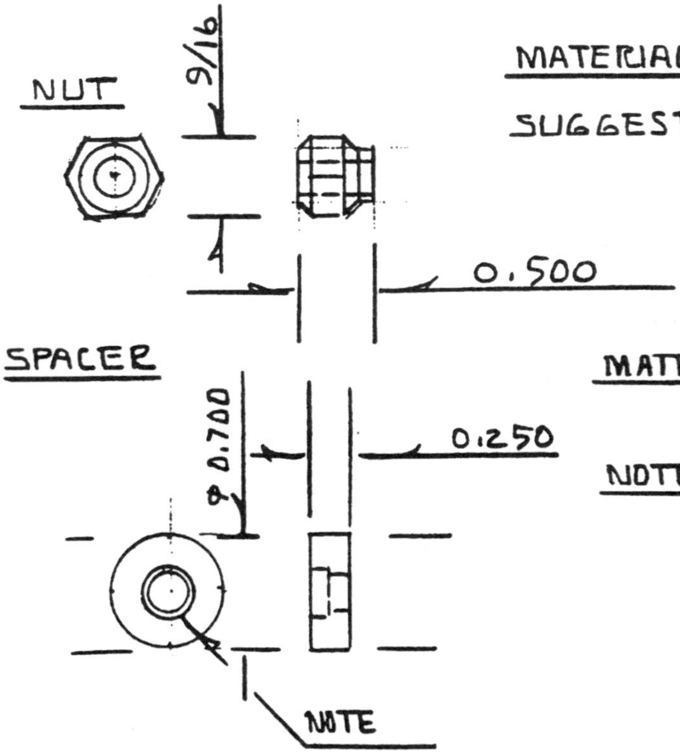

MATERIAL: 3/4 INCH BRASS ROD STOCK

NOTE: BORE 1/2-DEPTH TO ⌀ 0.335 - REMAINING DEPTH TO ⌀ 0.255

SHEET 5 OF 10

INDEX FIXTURE– INDEXING PLATES

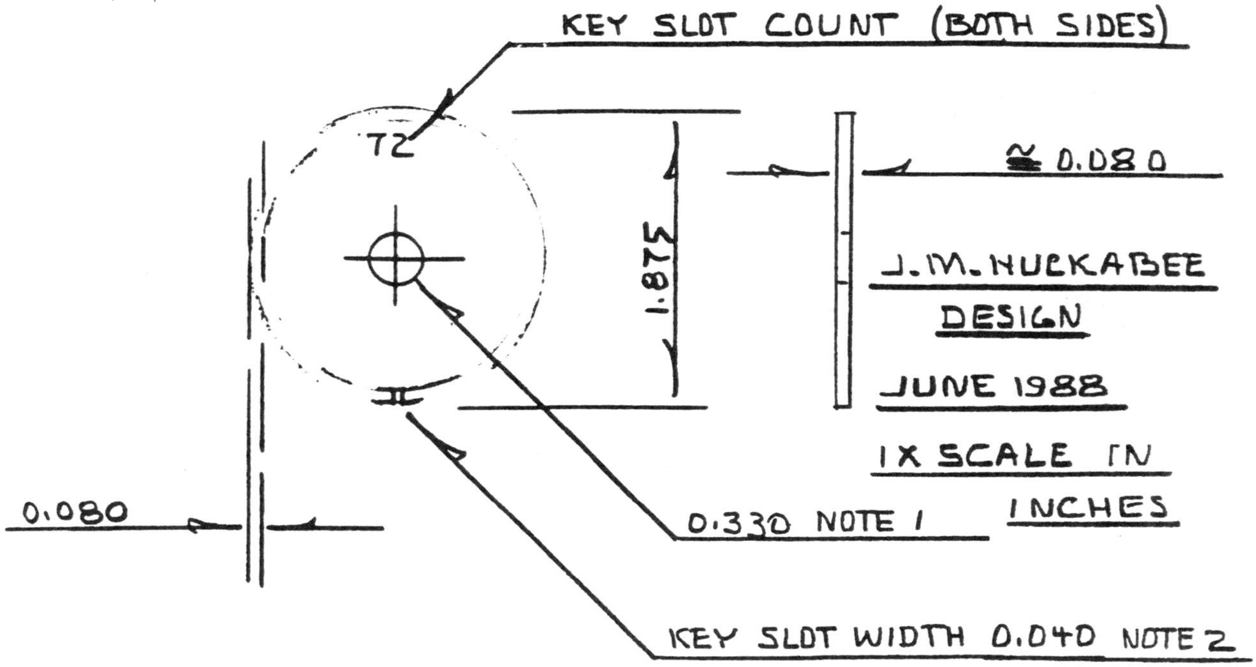

MATERIAL: BRASS OR HARD ALUMINUM

NOTE 1: THIS BORE MUST BE CONCENTRIC TO OD AND KEY SLOT ROOT DIAMETER CIRCLE; AND A ZERO-CLEARANCE FIT TO THE 0.330 SHOULDER OF THE WORK-ARBOR

NOTE 2: THIS KEY SLOT MUST BE A ZERO CLEARANCE WITH THE INDEX PIN

NOTE 3: CLOCKMAKERS WILL FIND KEY SLOT COUNTS IN THE RANGE OF 16–96 MOST USEFUL. A MODEST SET OF ABOUT 20 COUNTS WILL COVER THE MOST COMMON WHEELS ENCOUNTERED

NOTE 4: INDEX FIXTURE OPERATING INSTRUCTIONS WILL SHOW HOW TO MAKE THE KEY SLOT DIVISIONS

SHEET 6 OF 10

INDEX FIXTURE- INDEX PIN

J.M. HUCKABEE DESIGN
JUNE 1988
1 X SCALE IN INCHES

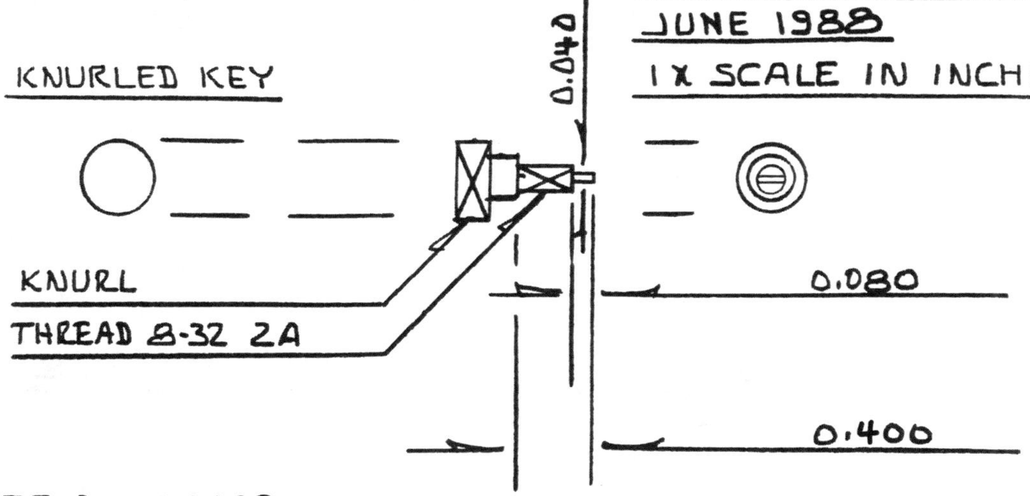

MATERIAL: BRASS

SUGGESTION: FINISH TIP AS DIMENSIONED, REMAINDER AS DEPICTED

INDEX FIXTURE-WHEEL BLANK SPACER

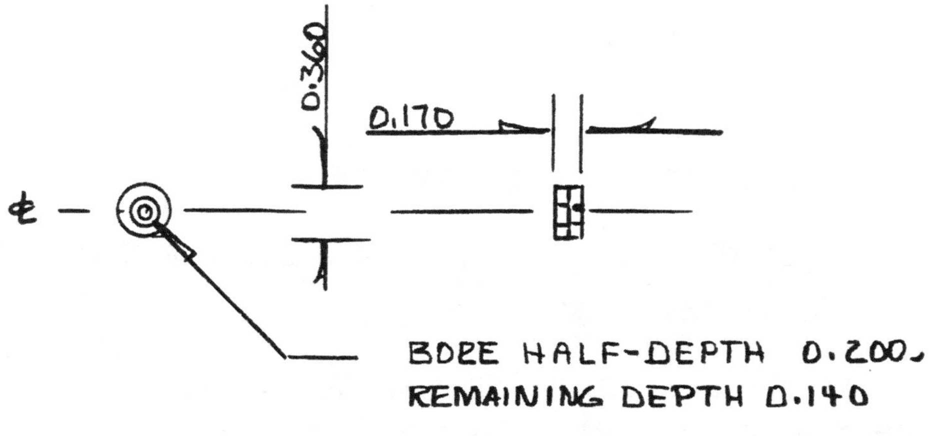

MATERIAL: 3/8 INCH BRASS BAR STOCK

SHEET 7 OF 10

INDEX FIXTURE- BACKUP PLATES

J.M. HUCKABEE DESIGN

JUNE 1988

1X SCALE IN INCHES

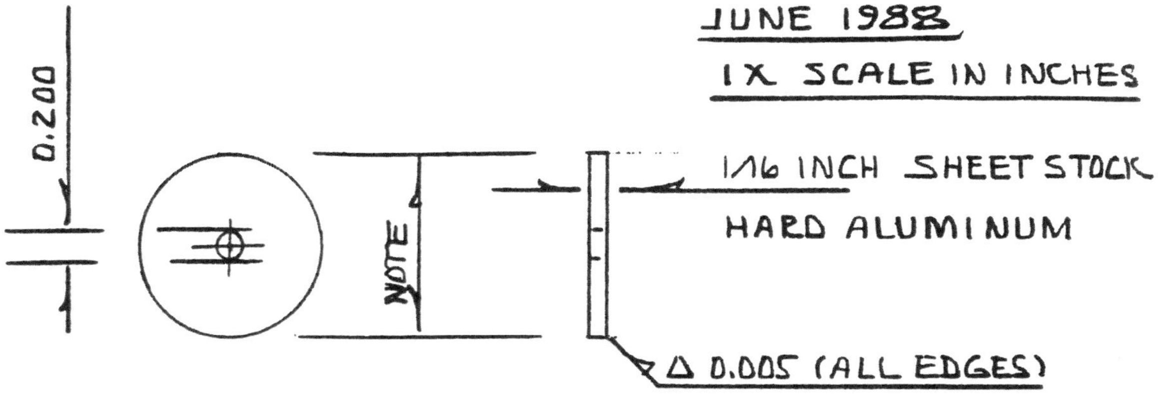

FINISH: WIRE BRUSH AND IMMEDIATELY WASH WITH HOT WATER & STRONG SOAP TO DEVELOP A SELF-OXIDE PROTECTIVE FINISH

NOTE: MAKE ABOUT 12 PAIRS OF PLATES WITH DIAMETER INCREMENTS OF ABOUT 1/8 INCH. BEGIN WITH DIAMETER OF 0.750 AND WORK UPWARD. THIS WILL COVER WHEEL DIAMETERS UP THROUGH ABOUT 2-1/4 INCHES

INDEX FIXTURE- HOLD-DOWN BOLT

THIS FIXTURE WAS DESIGNED FOR USE WITH A UNIMAT LATHE AND ITS 6MM BOLT & T-NUTS. A SOCKET-HEAD BOLT ≅ 1-1/2 INCHES LONG IS USED. A SUITABLE ALLEN WRENCH MUST BE MODIFIED (SHORT-END CUT OFF) IN ORDER TO PASS BETWEEN THE FIXTURE ARBOR AND THE BOLT HEAD: MODIFY BY GRINDING - AVOID DRAWING THE TEMPER. A SPECIAL BOLT MAY BE REQUIRED WHEN THE FIXTURE IS USED WITH OTHER LATHES.

SHEET 8 OF 10

INDEX FIXTURE- LATHE ARBOR

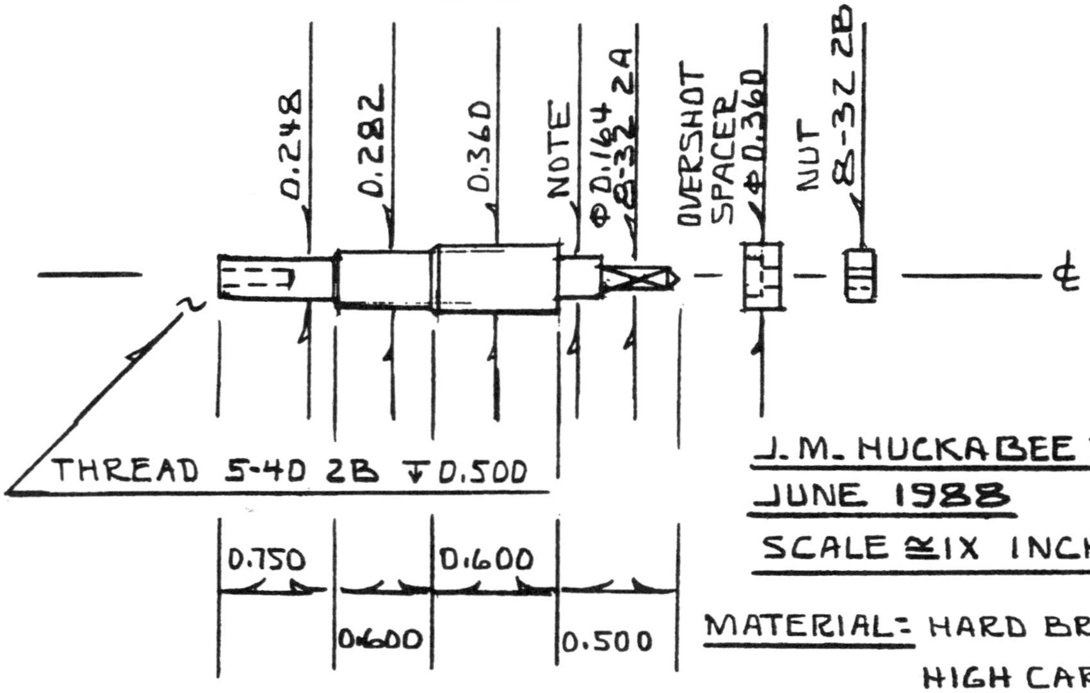

J. M. HUCKABEE DESIGN
JUNE 1988
SCALE ≅1X INCHES

MATERIAL: HARD BRONZE, HIGH CARBON STEEL, ETC.

INDEX FIXTURE - DRAWBOLT

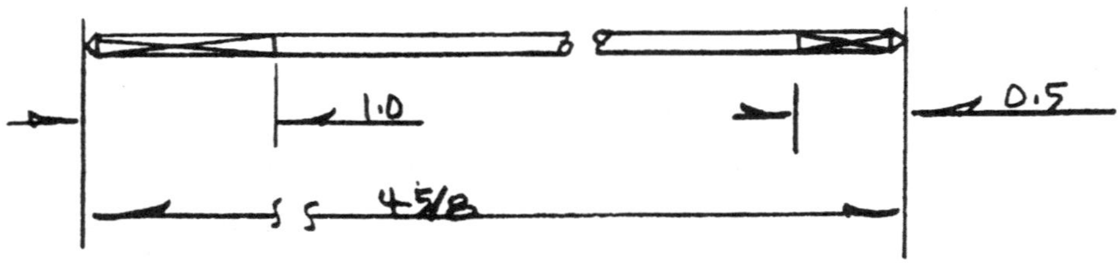

THREAD 5-40 2A (2X)

MATERIAL: COMMERICAL BRAZING ROD 1/8 INCH ⌀

SHEET 9 OF 10

INDEX FIXTURE - DRAWNUT

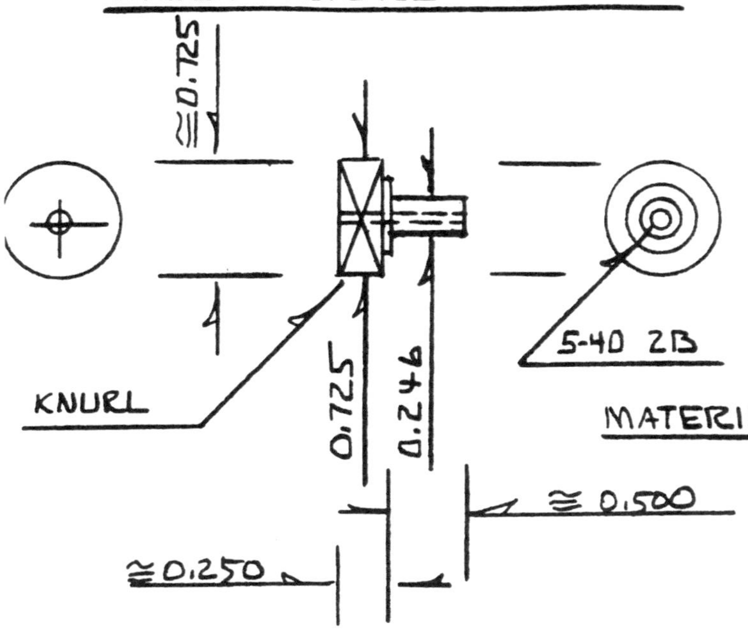

J. M. HUCKABEE DESIGN

JUNE 1988

1X SCALE IN INCHES

MATERIAL: 3/4 INCH BRASS ROD STOCK

DISCUSSION:

LATHE ARBOR, DRAW BOLT & DRAW NUT DIMENSIONS ARE DICTATED BY SPINDLE DIMENSIONS IN THE LATHE AT HAND. DIMENSIONS SHOWN ARE FOR AN OLDER UNIMAT.

ARBOR CUTTER SEATS AND CENTERS SHOULD BE FINISHED IN THE LATHE AT HAND, TO ASSURE THE HIGHEST ACCURACY. A SMALL PRICK-MARK ON THE ARBOR & LATHE SPINDLE HELPS IN FUTURE ACCURACY (LOCATOR).

MANY SMALL CIRCLE SAWS HAVE 0.250 BORE, WHILE MANY GEAR CUTTERS HAVE 7.0 MM BORE. TWO, OR MORE, ARBORS WILL BE REQUIRED, BUT ONLY ONE DRAW NUT & BOLT. OTHER THREAD SIZES MAY BE MORE READILY AVAILABLE

SHEET 10 OF 10

INDEX FIXTURE - COMMERCIAL ITEMS

THE FOLLOWING ITEMS ARE PURCHASED:

2- 6-32 X 1 RH SCREWS - PLATED
2- #6 X 3/8 FLAT WASHERS - PLATED
2- 6-32 X 3/8 RH SCREWS - PLATED
1- 6-32 X 1/4 NUT - PLATED
3- 6-32 X 1/4 HEADLESS SET SCREWS - PLATED
1- ALLEN, OR SPLINED, WRENCH FOR ABOVE
1- SOCKET-HEAD BOLT, WRENCH NUT, ETC., IF FIXTURE IS USED ON A LATHE OTHER THAN UNIMAT
1- CIRCLE SAW - APPROX 1 TO 1-1/2 INCH DIAMETER TO CUT INDEX PLATES & ESCAPEWHEELS. THICKNESS 0.040 W/BORE Φ 0.250
1- PROFILE GEAR CUTTER SUITABLE FOR THE JOB AT HAND: PREFERED DIAMETER IS 3/4 TO 1-1/4 WITH 7.0 MM BORE.
3- 8-32 X 11/32 NUT - PLATED

INDEX FIXTURE - OTHER CONSTRUCTION WORK

A SUITABLE INDEX PLATE MUST BE PREPARED: THIS WORK WILL BE SHOWN BY A VIDEO TAPE ALONG WITH SET-UP & USE OF THE FIXTURE

J. M. HUCKABEE DESIGN
JUNE 1988

APPENDIX B
ARCHIE PERKINS ON MAKING THE FLY CUTTER

The following is an excerpt reprinted with the kind permission of Archie Perkins, the author, and AWCI, the publisher of his series "Antique Watch Restoration" in the magazine *Horological Times*. The material appeared in the August 1989 issue. The instructions should be followed in the making of the fly cutters for the movement.

...from "Antique Watch Restoration", Part XLIV

When restoring antique watches and clocks, it is necessary to be able to make replacement wheels and pinions. Although it is possible to purchase some standard cutters for making some wheels and pinions, there are many times when the wheel or pinion needing duplication is not of the standard form and an exact cutter must be made in order to duplicate the wheel or pinion. In this case, the restorer must be prepared to make the cutter as well as to make the gear.

One of the most simple cutters to make is the fly cutter (or single tooth cutter). Fly cutters are best used for cutting brass wheels and pinions, although, if made strong and hard enough, they will also cut steel wheels and pinions.

Since a fly cutter has a single tooth, it must be turned ten times faster than a 10 tooth circular cutter in order to do the same amount of work. In other words, a single tooth cutter will remove one-tenth of the material that a 10 tooth cutter will remove per revolution of the cutter. When cutting steel wheels or pinions, a fly cutter will not hold up at great speeds unless the cutter is made of high speed steel or carbide. When cutting brass wheels and pinions, a fly cutter is used at high speeds but, when using a cutter made of high carbon steel for cutting steel wheels and pinions, the cutter must be turned at a very low speed with a very slow feed in order to protect the cutter from being damaged.

Figures 1 and 2 show the steps used for making a fly cutter to be used for cutting brass wheels and pinions. Note: A fly cutter made in this manner is not strong enough for cutting steel wheels and pinions, especially in the smaller sizes. This is due to the delicacy of its construction.

Figure 1 shows the first steps used in making the fly cutter. A piece of high carbon steel drill rod is used to make the cutter. The diameter of the rod should be a predetermined diameter and should be a size that is easily obtained because cutters made in the future should be made the same diameter to fit the same arbor chuck. The diameter of the rod should be 1/8 inch or 5/32 inch. To make a cutter, first chuck a piece of the rod true in the lathe. The length of the piece should be 3/4 inch. The first step used in making the fly cutter is to turn the end of the rod to fit the space between two good teeth on the damaged wheel. This is shown in View A, Figure 1. The graver used for this can be a diamond shaped graver which has had its point rounded to about the same radius as the radius of the tooth addendum. A graver can be made for this out of round high carbon steel drill rod. The rod should be of such a diameter that when its end is cut off at a 30 degree or 45 degree angle the point will have the proper radius for cutting the curve on the shoulder of the cutter. By varying the angle on the end of the round graver, different radii can be cut with its point. After the round graver has been formed at the point, then it is hardened and tempered to a very pale straw color. If the fly cutter is to cut a tooth with a square bottom then the end of the cutter is turned flat and square. On the other hand, if the cutter is to cut a round bottom, then the end of the cutter is rounded as shown in View B, Figure 1.

View C, Figure 1 shows the rod turned to fit the space between two good teeth. After this, fine polishing paper should be used on the turned section to produce a smooth polished surface.

The next steps used to complete the forming of the cutter are shown in Figure 2. View A shows how one side of the turned section is filed flat in order to form the front face of the cutter and the cutting edges on the cutter. This surface should be filed just to the center of the rod. If this surface is filed past the center of the rod, the cutter will be too narrow and will cut a narrow space which will leave the teeth on the wheel that is cut with the cutter too thick. On the other hand, if the surface is not filed to the center, the cutter will not have natural relief and the Fig-

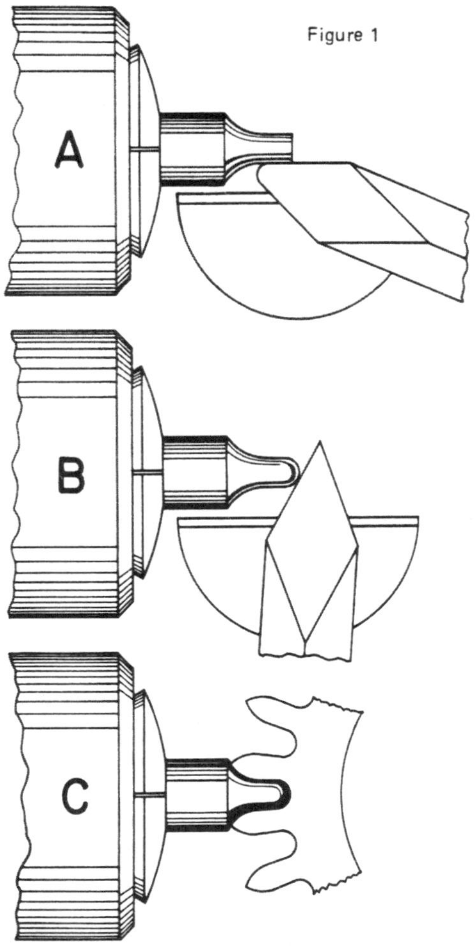

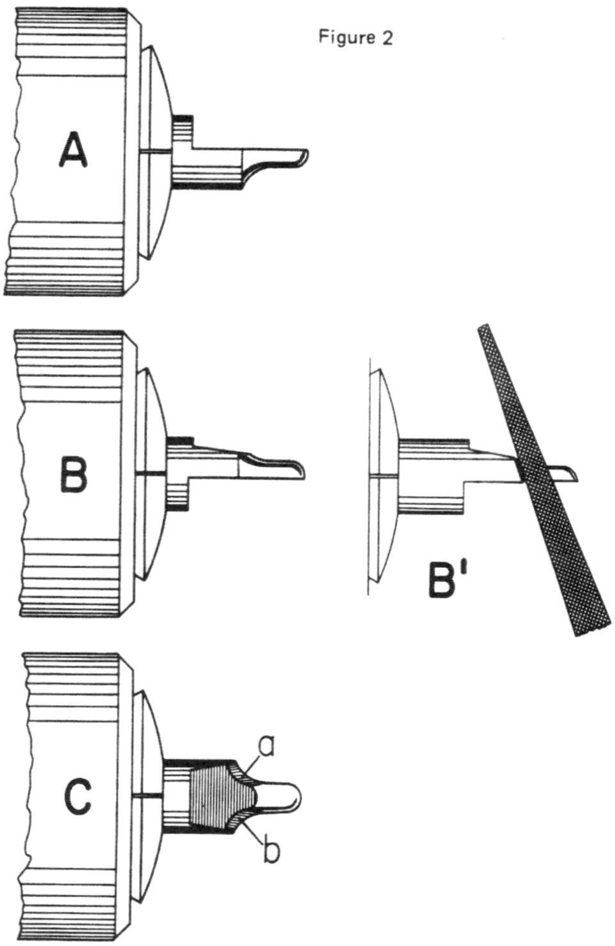

ure 1 cutter will drag and not cut well. When filing the flat surface, the headstock is locked up with the locking pin for the headstock.

The next step is to file the back side of the cutter as shown in View B, Figure 2. This is done so the curved shoulder of the cutter can be relieved more easily. This surface should be filed flat but sloped slightly toward the point of the cutter.

The next step is to file the relief on the curved shoulder of the cutter. This is done with a fine tapered round needle or escapement file. This is shown being done in View B, Figure 2, in the right-hand image. A section of the tapered round file is used which has the same radius as the curve on the shoulder of the cutter that cuts the curved addenda of the tooth. The curved relief on the cutter is filed until the relief surface **almost** reaches the flat face of the cutter. The file should never cut into the flat face of the cutter; if this hap pens, the shape of the cutter will be changed and, as a result, the cutter may not cut a tooth which is shaped correctly. A small land should be left just below the flat face of the cutter to allow for grinding the flat face of the cutter to sharpen it after the cutter has been hardened and tempered. The curved shoulders on both sides of the flat face of the cutter must be relieved in the same manner with the file.

View C, Figure 2 shows the fly cutter after it has been relieved with the file at points “a” and “b”. If the cutter has a rounded point, then the sides and end of the cutter will have a natural relief when the cutter has been ground to the center of the stock. If the cutter has a square end, then the end of the cutter must be relieved by grinding the end of the cutter underneath the flat face of the cutter at about a 10 degree angle.

After the cutter has been relieved, then it is hardened and tempered to a pale straw color. The shank of the cutter should not be hardened. This assures that the set screw of the arbor chuck will bite into the cutter shank to hold the cutter tightly in its cross hole in the arbor chuck when the cutter is being used. In other words, if the cutter shank is too hard for the set screw to bite into, then the cutter will work loose in its hole in the arbor chuck while cutting a wheel.

HARDENING THE CUTTER

To harden the cutter, it is first heated slightly, then placed in powdered boric acid so a light uniform coating of the powdered boric acid adheres to the surface of the cutter. If the cutter is not heated enough, the powder will not stick to the cutter. On the other hand, if the cutter is heated too hot, the powder will lump up on the cutter and be difficult to remove from the cutter after the cutter has been hardened. The powdered boric acid is an anti-oxidizer which keeps the surface of the steel clean during the hardening process. The powdered boric acid coating sheds off of the steel when it is quenched during hardening. After the coating of powdered boric acid has been applied to the surface of the cutter, then the cutter is reheated to a uniform cherry red and then quickly quenched, end first, into a container of water. The water should be at room temperature. Note: During this process, the shank of the cutter is held with a soldering tweezer or an old pair of pliers or tongs while the cutting end of the cutter is heated. The end of the cutter is fragile and must not be overheated. This could harm the cutter and make it very brittle and subject to breaking even after being tempered. When a piece of steel has been overheated and quenched, the grain in the steel becomes very coarse. This damages the steel and makes it weak. When a piece of steel is not heated enough for hardening, it can be reheated to harden it, but if the steel is overheated, it has permanent damage.

After the cutter has been heated and quenched, then it is checked for hardness. This is done by rubbing an old fine file over the face of the cutter. The file should slide over the surface freely without biting into the surface. If the file should bite into the surface, then the cutter has not been sufficiently hardened. In this case, the cutter must be reheated slightly hotter than previously heated and re-quenched in the water. Cold tap water will cause the steel to harden to a harder state than does water at room temperature. Also, the cutter must be quenched quickly after reaching the desired color and not allowed to cool any before being quenched.

TEMPERING THE CUTTER

After the cutter has been sufficiently hardened, then its surface is cleaned off with emery paper to prepare the cutter for tempering. The cutter is held in the lathe while the emery paper is used to clean off any scale from its surface. Note: if the cutter is to cut steel, it should not be tempered but should be left dead hard. If the cutter is to be used to cut brass gears, it should be tempered to a very pale straw color. The tempering can be done by placing the cutter on top of some clean white sand in a metal pan which is heated slowly over an alcohol flame. The cutting section of the cutter is closely watched and, when that section reaches the desired color, the cutter is removed from the sand and placed on a slab of glass to cool the cutter gradually.

Another method that can be used for tempering the cutter is to heat the end of the cutter shank in an alcohol flame until the cutting end reaches the desired color, then dip the cutter into water to stop the process. The cutter is held by the cutting end with the soldering tweezer while being tempered.

CLOCK BUILDER'S NOTES